4-10-95

Practical Electron
Microscopy

D1260467

Practical Electron Microscopy

A beginner's illustrated guide
Second edition

Elaine Hunter
University Hospital, London, Ontario

With contributions from
Peter Maloney and Moïse Bendayan

CAMBRIDGE
UNIVERSITY PRESS

Published by the Press Syndicate of the University of Cambridge
The Pitt Building, Trumpington Street, Cambridge CB2 1RP
40 West 20th Street, New York, NY 10011-4211, USA
10 Stamford Road, Oakleigh, Victoria 3166, Australia

First edition published by Praeger Publishers 1984

This edition published 1993

Printed in the Canada

Library of Congress Cataloging-in-Publication Data
Hunter, Elaine E. (Elaine Evelyn)
Practical electron microscopy : a beginner's illustrated guide /
Elaine Hunter ; with contributions from Peter Maloney and Moïse
Bendayan. – 2nd ed.
p.   cm.
Includes bibliographical references and index.
ISBN 0-521-38539-3 (pbk.)
1. Electron microscopy – Technique.   2. Transmission electron
microscopes – Laboratory manuals.   I. Maloney, Peter, 1955–
II. Bendayan, Moïse.   III. Title.
[DNLM: 1. Microscopy, Electron – methods.   QH 212.E4 H945p]
QH0212.E4H86   1993
578'.45–dc20
DNLM/DLC   92-49456
for Library of Congress   CIP

A catalog record for this book is available from the British Library.

ISBN 0-521-38539-3 paperback

To my husband,
Iain

• *Contents*

• *Foreword*

This is a well-written, beautifully illustrated manual for electron microscopy. It reflects Elaine Hunter's very considerable experience in this field and offers both those setting out to use electron microscopic techniques and experienced individuals very useful information.

In tradition, this book is related to *Electron Microscopy, A Handbook for Biologists* written by Edgar Mercer, who was in charge of the Electron Microscopy suite at the John Curtin School of Medical Research at the Australian National University during my sojourn there as a postdoctoral fellow and Daniel Pease's *Histological Techniques for Electron Microscopy* used when I established an electron microscopy laboratory in Toronto in the 1960s. In Ms. Hunter's book the reader is taken through chapters on the handling of tissues and the necessary steps in fixation, processing, embedding, and examination of tissue to assure the best results from electron microscopy. I particularly like the comparative photographs with which the author illustrates the use of different techniques in tissue preparation and believe the "trouble-shooting" guide, when using an electron microscope, most useful. Special techniques commonly used in electron microscopy are covered and details of the importance of recording observations photographically are given. Above all, the importance of keeping detailed records of all activities is emphasized. Information related to protection of the health of those engaged in electron microscopy and for the disposal of noxious agents used is not forgotten.

Elaine Hunter is to be congratulated, not only in extending tradition but, in this book, making a reader aware of technology in this field.

M. D. Silver, Chairman
Department of Pathology
University of Toronto

• *Preface*

This book is an illustrated workbench manual of electron microscopy techniques. All the methods cited have been used successfully by the authors in both clinical and research laboratories and have proven to be both reliable and reproducible.

It is intended that someone with very little experience and using only this manual as a guide could set up and operate an electron microscopy laboratory. The preparative techniques outlined apply to animals and human material and are not necessarily applicable to plants and insects. Some will be especially helpful in a clinical setting, which is where the author is presently employed.

Usually only one technique has been outlined in each section, and the methods and materials are very specific and detailed. This has not been done to discredit other methods or products. Rather, it has been done for simplicity in order to eliminate pitfalls for the beginner, to get a reliable method working quickly and easily, and to present as many "tricks of the trade" as possible. Although the methods outlined have been tested extensively and are reliable, scrupulous attention to detail when both reading and applying them is advised. Laboratory cleanliness and accuracy in weighing and mixing chemicals are extremely important. Many of the chemicals used in electron microscopy are very dangerous, and great care should be exercised both in the handling and disposal of these chemicals.

References and descriptions have been kept to a minimum for the sake of clarity and to avoid confusing the novice. There are several very good texts on the market that go into great theoretical detail and provide a variety of special methods once the technologist has gained experience.

It was decided to include a section on immunoelectron microscopy and Dr. Moïse Bendayan, who is a leading expert in the field, very kindly consented to write this chapter. While this is not a field for the beginner, it is an area that is gaining popularity, and most laboratories in the biological field require personnel with a knowledge of the subject. Although it is a complicated field, Dr. Bendayan has made every effort to outline as simply as possible some basic techniques. Experience, expertise, and a wealth of current literature (much of it written by Dr. Bendayan) will enable the technologist to develop the area once basic skills have been mastered. I wish to thank him for his excellent contribution.

The chapter on the electron microscope was contributed by Peter Maloney who has been an electronic equipment specialist with Philips Electronics for ten years. In addition to being an expert service person and excellent trouble shooter, Peter has taken a complicated subject and made it both interesting to read and easy to understand. The line drawings were done by Peter's brother, John, who is a civil draftsman. The drawings and the picture of the transmission electron microscope were used with the permission of Philips Electronics.

Few authors have had the advantage of the dedicated and meticulous editorial assistance given by my husband, Iain. He has devoted hours to reading, changing, and sometimes rewriting with the intended audience in mind. After years of doing the same techniques, it is easy to forget the many little traps one fell into as a beginner. Iain never failed to ask: Why? How? Why that way and not this way? Did you really mean to say this? Are you sure you didn't leave something out here? I thank him for his patience, guidance, and help.

Most of the manuscript was typed into WordPerfect by Mr. Geoffrey Thoms, Computer Science teacher at Lord Dorchester Secondary School, Middlesex County, Ontario. His help was greatly appreciated. I am grateful also to University Hospital for providing such excellent facilities. In particular my thanks go to Peter Munavich of the Pathology Department, to Kathy Stuart and Steve Mesjaric, Instructional Resources, for their photographic assistance, as well as to George Moogk for many of the line drawings. All are employed at University Hospital.

Special thanks are due to Cambridge University Press and especially to Dr. Robin Smith, whose patience and support through some rather difficult times went well beyond the call of duty.

Elaine Evelyn Hunter
London, Ontario
Canada

1

• *Fixation*

INTRODUCTION

The importance of good fixation in electron microscopic (EM) studies cannot be overemphasized. The purpose of fixation is to preserve tissue structure with minimal alteration during dehydration, embedding, cutting, staining, and viewing in the electron microscope. Since important chemical bonds in living tissue depend upon the presence of water for their stability, a fixative should provide the stable bonds necessary to hold the molecules together during dehydration. Fixatives form cross-links, not only between their reactive groups and the reactive groups of the tissue but also between different reactive groups within the tissue (Hayat, 1986, pp 1–2). A poorly fixed specimen, or one that has dried even slightly prior to fixation, does not dehydrate properly or infiltrate evenly with the embedding medium. This results in sections that fracture during cutting (Fig. 1.1A), disintegrate in the water bath, stain unevenly (Fig. 1.1B), or become severely disrupted by the electron beam (Fig. 1.1C).

The most common reason for poor fixation is *large specimen size*. For example, glutaraldehyde, the fixative most often used in electron microscopy, penetrates to a depth of less than 1 mm. To minimize autolytic changes, slices or ribbons of tissue 0.5 mm thick should be placed in fixative promptly. Thicker slices, which are subsequently reduced in size in the laboratory, invariably result in poor-quality fixation.

Coagulant fixatives (e.g., alcohol) transform proteins into opaque mixtures of granular or reticular solids suspended in fluid. This causes considerable distortion of fine structural detail, making the tissue unsuitable for electron microscopy.

Noncoagulant fixatives transform proteins into transparent gels, thus stabilizing them without much structural distortion. Osmium tetroxide, glutaraldehyde, acrolein, and formaldehyde are all noncoagulant fixatives. Osmium and ruthenium tetroxides are also considered *additive fixatives* since they bond to the proteins they fix. Also, because they are salts of heavy metals, they impart density to the tissue.

The most common and best method of fixation for electron microscopy involves the use of glutaraldehyde followed by osmium tetroxide (Fig. 1.3A). Acrolein, potassium permanganate, and certain electron-dense stains such as uranium salts and ruthenium tetroxide are also used as fixatives.

Osmium tetroxide (OsO_4) acts as a fixative as well as an electron-dense stain. In its reduced state, this fixative renders tissue black, permitting it to be more easily seen during processing and cutting. Osmium tetroxide penetrates slowly but reacts quickly

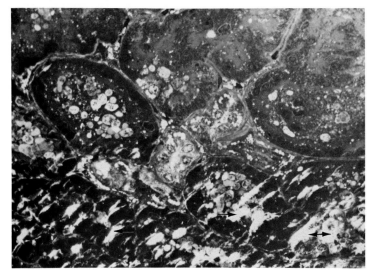

Figure 1.1. *Effects of poor fixation.*

A. Light micrograph showing fractures (*arrows*) in a 1-μm section of liver. This is due to poor processing, a direct result of bad fixation. Specimen is fragile, will produce poor thin sections, and will disrupt easily under the electron beam.

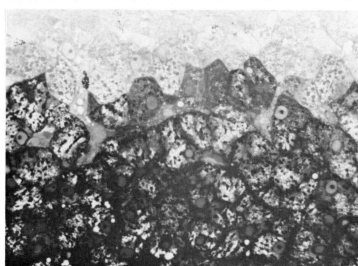

B. Light micrograph of liver showing uneven staining as well as some fracturing of a 1-μm toluidine blue section.

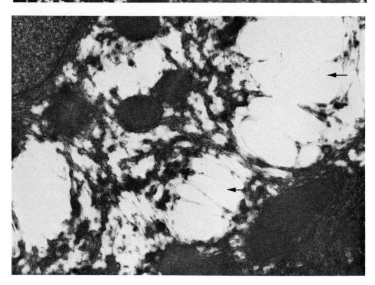

C. Electron micrograph of the same specimen shown in Fig. 1.1A. Exposure to the electron beam has caused large holes (*arrows*) to form in the section.

with tissue. This means that the fixed outer layers impede progress of the fixative, resulting in a penetration of less than 1 mm. It follows that blocks less than 1 mm thick are recommended if fixation is to be uniform. Osmium tetroxide fixes lipids, proteins, and lipoproteins. It does not react with ribonucleic acid (RNA) or deoxyribonucleic acid (DNA) and, in low concentrations, does not preserve microtubules. Prolonged fixation in osmium tetroxide will extract some proteins. Osmium tetroxide that has been reduced by potassium ferrocyanide (Karnovsky, 1971), while not employed in the methods described in this chapter, is used in many laboratories. It delineates membranes well and gives enhanced staining of glycogen (Fig. 1.3B) and the extracellular matrix of bone and cartilage (Lewison, 1989).

Ruthenium tetroxide (RuO_4) is closely related to osmium tetroxide. It decomposes very readily, even in a refrigerator, and is then useless as a fixative. It delineates the triple-layered structure of cell membranes and reacts strongly with proteins, glycogen, and monosaccharides. It penetrates more slowly than osmium tetroxide.

Potassium permanganate ($KMnO_4$) penetrates somewhat further than osmium tetroxide. Much cellular material is lost during and after potassium permanganate fixation, but the high contrast achieved in cell membranes makes it a valuable fixative for the study of these structures. It preserves DNA and prevents its clumping during dehydration. Potassium permanganate, osmium tetroxide, and ruthenium tetroxide are strong oxidizing agents, which can destroy enzymatic activity in cells. This gives them limited value as fixatives in cytochemistry. Potassium permanganate does not preserve lipids.

Glutaraldehyde ($C_5H_8O_2$) is a dialdehyde and the most effective of the aldehyde fixatives for preserving fine structure (Sabatini, Bensch, & Barrnett, 1962; Sabatini, Bensch, & Barrnett, 1963; Sabatini, Miller, & Barrnett, 1964; Barrnett, Perney, & Hagström, 1964). It stabilizes cell structures, prevents distortion during processing, and increases the permeability of tissue to embedding media. No other fixative surpasses glutaraldehyde in its ability to cross-link proteins (Weakley, 1981, p. 25). It does not impart contrast to the tissue nor make lipids insoluble; in fact, these tend to leach out and form myelin membranes, which, when later exposed to osmium tetroxide, are the "myelin figures" of poorly fixed material (Ghadially, 1980) (Fig. 1.2). Because of the inability of glutaraldehyde to stabilize lipids, cell membranes will not be visible unless tissue is later postfixed in osmium tetroxide (Fig. 1.3C). Therefore, postfixation in osmium tetroxide is needed to lend contrast and to further stabilize fine structure so that it can withstand dehydration and embedding in plastics. If white blood cells are to be examined, glutaraldehyde–osmium tetroxide (mixed 1:1 immediately before use) fixes better than when each is used separately (Hayat, 1986, p. 4; Hayat, 1989, p. 56).

Acrolein (C_3H_4O) (also known as acrylic aldehyde) is a monoaldehyde. It penetrates and reacts faster than most fixatives and causes little shrinkage. It can be used when relatively large pieces of tissue must be taken for electron microscopy. Acrolein and glutaraldehyde, when used together, give better preservation of microtubules than either aldehyde would do on its own. Acrolein has a tendency to polymerize and generate heat when exposed to light, air, or certain chemicals. The commercial product contains an oxidation inhibitor. Acrolein is an extremely hazardous chemical, being inflammable, toxic through inhalation or skin absorption, and strongly irritating to the skin.

Formaldehyde (HCHO) is also a monoaldehyde. As commercial formalin (37% gas

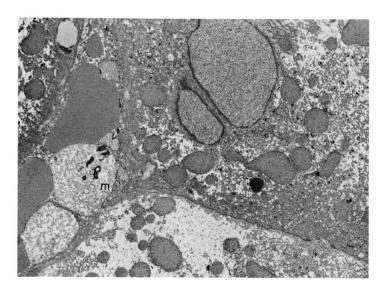

Figure 1.2. Electron micrograph of poorly fixed liver tissue. Myelin figures (m) due to poor fixation are visible. Intracellular organelles are also very disrupted.

in water), it is the most commonly used fixative for light microscopy. It is sometimes used for electron microscopy but is then best prepared from paraformaldehyde powder. It is useful for fixing very dense tissue, which would not be penetrated by glutaraldehyde. It forms weak cross-links and is therefore not as good a fixative for fine structure as glutaraldehyde but can be used quite successfully for diagnostic purposes (Fig. 1.3D) (Ghadially, 1985). Formaldehyde penetrates more rapidly than glutaraldehyde (having a smaller molecular size) but fixes more slowly. It reacts with proteins, lipids, and nucleic acids and can be combined with glutaraldehyde to form a "universal" fixative suitable for both light and electron microscopy (McDowell & Trump, 1976).

Uranium salts act both as electron-dense stains and as sequential fixatives. If used prior to dehydration, they help to stabilize fine structures and prevent clumping of DNA. They impart contrast to tissues whether used prior to or following dehydration, but, if used prior to dehydration, glycogen will be extracted (Bullock, 1984, p. 4).

The *pH of a fixative* should be close to the pH of the tissue to be fixed; otherwise, the structure and behavior of proteins will be altered. Buffers are present in living cells, and changes in fixative pH may cause intracellular changes such as alterations in the consistency of the matrix, permeability of cell membranes, and activity of enzymes. Since the pH of most animal tissues is 7.4, the most satisfactory preservation of fine structure is obtained by keeping the pH of the fixative between 7.2 and 7.4. There are exceptions; for example, plant cells fix best at pH 8.0 and gastric mucosa at pH 8.5 (Hayat, 1989, p. 14). The optimum pH for any specific tissue may be determined experimentally, but in a routine service laboratory where many types of tissue are under investigation, one must "strike a happy medium" (pH 7.2–7.4).

Some thought should be given to the chemical composition of the buffer. This is more critical when buffering osmium tetroxide than glutaraldehyde, since the latter forms stable cross-links with the tissue components. When deciding upon a buffer, the type of tissue to be studied must be the determining factor. Osmium tetroxide buffered with *s*-collidine, for example, extracts more proteins from rat liver than any of seven other buffers and yet is best for fixing lung tissue (Luft & Wood, 1963). Also, since phosphate buffers are found in living systems, one might consider these the

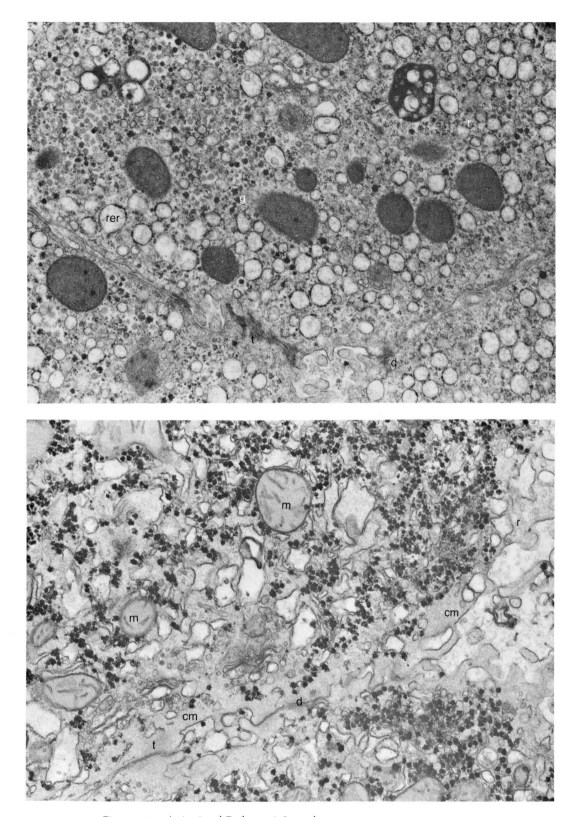

Figure 1.3. **A.** (top) and **B.** (bottom). Legend on p. 7.

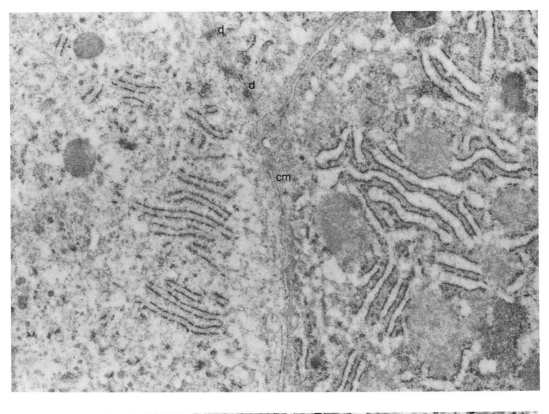

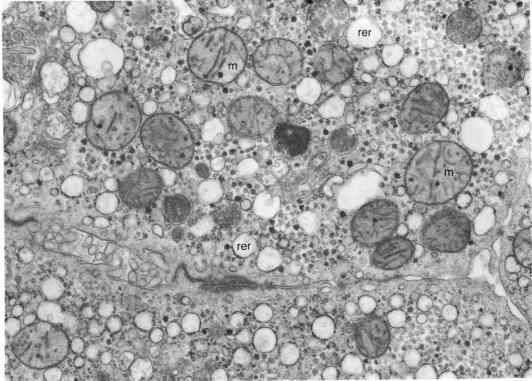

Figure 1.3. C. (top) and **D.** (bottom). Legend on p. 7.

best to use. In fact, they cause artifacts in the form of electron-dense spherical granules in some tissues (Gil & Weibel, 1968; Bullock, 1984, pp. 7, 8). Some of the buffers used most commonly are cacodylate, s-collidine, phosphate, and veronal acetate.

Another factor to consider when preparing fixatives is *tonicity*. If a fixative is isotonic, it may penetrate too slowly; if it is hypertonic, it will cause tissue shrinkage; if it is hypotonic, it will cause tissue swelling. Control of these changes is especially important for those doing morphometry or stereology. In practice the best results are obtained by using a fixative that is slightly hypertonic. Often nonelectrolytes such as sucrose or electrolytes such as sodium chloride are added to adjust the tonicity.

Temperature of fixation is also important. One must strike a balance between speed of fixative penetration, which is slow at low temperatures, and autolytic changes, which occur more rapidly at higher temperatures. These factors vary according to the type of fixative used and the type of tissue to be fixed. Fixation at room temperature immediately upon removal from the host is satisfactory in most situations. Fixative is usually stored at 4°C and used either at this or room temperature.

Duration of fixation must be considered. Once again, this varies with the type of tissue, the size of the block, the temperature, the type of fixative, and the buffer. Prolonged fixation extracts cellular materials. This is more likely to be a problem with osmium tetroxide than with glutaraldehyde, since the former does not cross-link many proteins, which can be extracted by the solvent while the tissue is being fixed. Thirty minutes to 1 hour is adequate in osmium tetroxide. The required minimum time in 2.5% glutaraldehyde is 1 hour and no benefit is gained beyond 2 hours.

Tissue specimen size is very important. Most poor fixation results from large specimen size and/or delay in getting the specimen into fixative. Tissue is fixed uniformly only if block size is kept small (Fig. 1.4). Osmium tetroxide fixes rapidly and uniformly to a depth of 0.25 mm; glutaraldehyde fixes to a depth of 0.5 to 1 mm. Therefore, the ideal thickness of a specimen should not exceed 0.5 mm for glutaraldehyde fixation (Hayat, 1989, p. 7). To avoid the mechanical damage caused by mincing fresh tissue, thin slivers can be cut, fixed, and later trimmed to 0.5-mm cubes. Should a large piece of fixed tissue be submitted, a very thin slice (less than 0.5 mm) must be shaved from the surface because deeper portions will be poorly fixed or not fixed at all.

The quality of fixation is not only affected by the characteristics of the fixative but also by the *method of fixation*. The best preservation is obtained by perfusing fixative through the blood vascular system of a previously anesthetized animal (in vivo). In situ fixation – flooding an area within the anesthetized animal with fixative, or micro-injecting fixative into the living tissue – also gives excellent results. The least effective method, but the one that must be used most often, is removal of tissue from the body and subsequent immersion in fixative (in vitro). This, of course, should be done

Figure 1.3. *Electron micrograph of tissue from a needle biopsy of human liver.*
   A. Tissue fixed in glutaraldehyde, post-fixed in OsO_4. Desmosomes (d) and tonofilaments (t) are clearly visible. Glycogen (g), ribosomes (r), and rough endoplasmic reticulum (rer) are also visible.
   B. Tissue fixed in glutaraldehyde and postfixed in potassium ferrocyanide-reduced osmium tetroxide. Membranes (cm) are well delineated, but there is a loss of mitochondrial matrix (m). Ribosomes (r), tonofilaments (t), and desmosomes (d) are barely visible.
   C. Tissue fixed in glutaraldehyde with no postfixation in OsO_4. Membranes (cm) have largely disappeared; desmosomes (d) are barely visible.
   D. Tissue fixed in formaldehyde and postfixed in OsO_4. Some mitochondrial matrix (m) is lost. Endoplasmic reticulum (rer) is more vacuolated than in A.

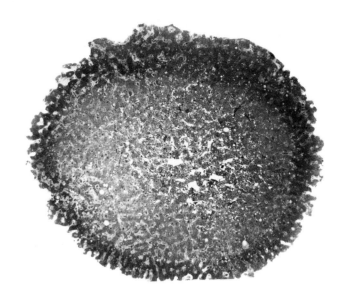

Figure 1.4.　Poor fixation due to large specimen size is shown in this light micrograph of a 1-μm section of liver stained with toluidine blue. Although this needle biopsy was only 1 mm thick, the fixative has not penetrated evenly to the center of the tissue.

quickly and with as little mechanical damage as possible. Since it is absolutely critical that the tissue not dry even slightly at this stage (Figs. 1.1 & 1.8), it may be stored temporarily in phosphate-buffered saline, if no fixative is available.

It is worth noting that, while the fixation procedure described later in this chapter gives excellent results for most routine EM material, it might require modification to suit specific tissues or projects in fields such as morphometry, immunocytochemistry, histochemistry, and elemental analysis. If fixation is performed carelessly, subsequent procedures are a waste of time and chemicals.

CRITERIA FOR GOOD ULTRASTRUCTURAL PRESERVATION

Plasma membrane	Intact and smooth around entire cell.
Cytoplasmic matrix	Fine precipitate. No empty spaces present.
Ribosomes	Clearly visible.
Mitochondria	External and internal membranes smooth. No gross distortion in shape. Very few swollen or empty looking.
Endoplasmic reticulum	Membranes intact and parallel. Stacked cisternae arrangement uniform.
Golgi apparatus	Vesicles of various sizes appear in a circumscribed area. Membranes intact.
Nuclear envelope	Both membranes intact (except at pores) and essentially parallel to each other.
Nuclear contents	Finely granular ground substance. Dense chromatin.

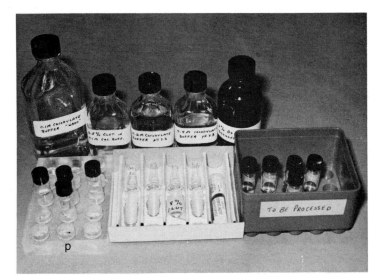

Figure 1.5. *Materials recommended for fixation.*

A. Glutaraldehyde and osmium tetroxide in sealed vials, prescription bottles, amber bottle (to protect osmium tetroxide from light), small specimen vials, and plexiglass (Perspex) support (p).

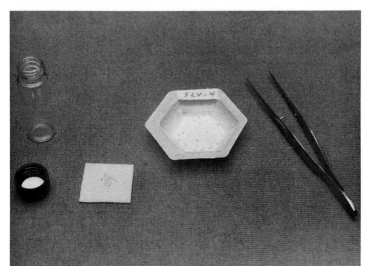

B. Iris forceps, specimen vial with Ker cap, dental wax supporting tissue covered with fixative, and small weighing boat containing tissue.

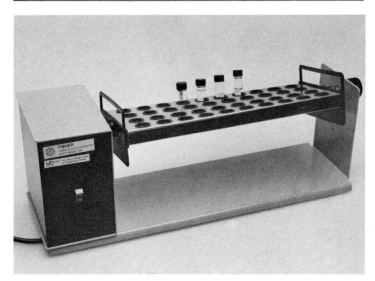

C. Rotator carrying specimen vials containing tissue and identification labels.

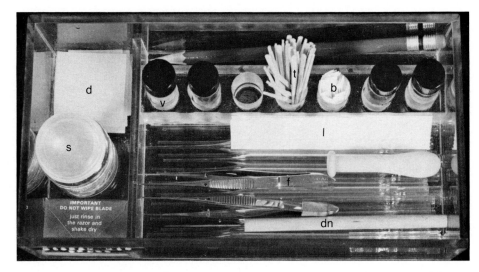

Figure 1.6.   Perspex specimen collection kit 215 × 114 × 63 mm (8.5 in. × 4.5 in. × 2.5 in.). (d) Squares of dental wax. (b) Strips of blotting paper. (s) Electrolyte solution. Tissue may be placed on blotting paper on a piece of dental wax and moistened with electrolyte solution for a few minutes if it cannot be fixed immediately. (l) Tissue identification labels. Lipshaw No. P-50 are perforated to tear conveniently and fit the specimen vials perfectly. (v) Specimen vials containing fixative. (t) Flat toothpicks. (f) Iris forceps. (dn) Dissecting needle.

CHEMICALS AND EQUIPMENT

- Glutaraldehyde − 8% aqueous in nitrogen-sealed vials or 25% EM grade (Fig. 1.5A).
- Osmium tetroxide − 1 g in sealed, prescored vials, or 4% aqueous in sealed vials (Fig. 1.5A).
- Sodium cacodylate − $(CH_3)_2AsO_2Na \cdot 3H_2O$.
- Hydrochloric acid (HCl), concentrated − 36.5%.
- Distilled water (or preferably double-distilled, deionized water).
- Formalin, commercial grade (37%−40% formaldehyde gas in water).
- Sodium biphosphate $(NaH_2PO_4 \cdot H_2O)$.
- Phosphate-buffered saline.
- Volumetric flask − 1 L.
- Prescription bottles − 50 and 100 mL with Ker caps (Fig. 1.5A) or 125- and 250-mL rectangular plastic bottles.
- Amber glass 50-mL bottles for osmium tetroxide (Fig. 1.5A).
- Specimen vials − 3.8 mL with Ker caps (smooth, white rubber liners. NOTE: Vinyl or soft rubber liners are dissolved by dehydrating or infiltrating fluids) (Fig. 1.5B).
- Graduated cylinders − 5, 10, 25, 50, 100, and 250 mL.
- Razor blades, double-edged − bulk-packed in boxes of 100.
- Pasteur pipettes − 145 mm (5 in.), rubber bulbs.
- Detergent (e.g., Teepol, Liquinox).
- Kitty litter.
- Corn oil.
- Dental wax (Figs. 1.5B & 1.6).
- Iris forceps − stainless steel (Fig. 1.5B).
- pH meter.

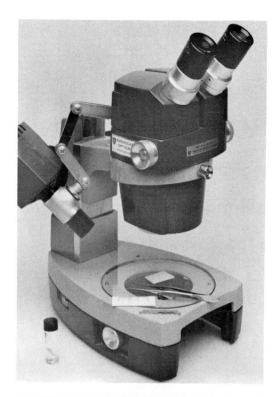

Figure 1.7. *Materials and equipment for trimming and fixing tissues.*
A. Dissecting microscope showing dental wax, halved razor blade, iris forceps, and millimeter ruler. A small, labeled specimen vial sits nearby.

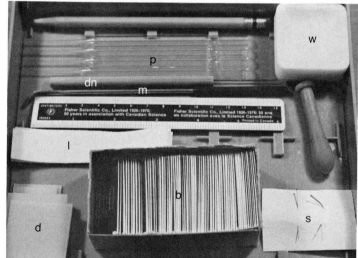

B. Convenient kit to keep beside the dissecting microscope to use when fixing and later, embedding tissue. Contains Pasteur pipettes (p), dissecting needle (dn), weighing boats (w), razor blades (b), dental wax (d), bent microspatula (m), labels (l), and pins (s). A box of specimen vials could also be kept nearby.

- pH paper, short-range – pH 6.8–8.5.
- Toothpicks, flat (Fig. 1.6).
- Labels in perforated sheets (e.g., Lipshaw P-50) – 7 × 20 mm (9/32 in. × 13/16 in.) (Fig. 1.6).
- Weighing boats – 35 mm (1.5 in.) (Fig. 1.5B).
- Perspex block, drilled to support specimen bottles (Fig. 1.5A).
- Laboratory film (Parafilm) – 102 mm × 3,810 cm (4 in. × 125 ft.) roll.
- Plastic bags – 2 L (6 lb).
- Dissecting microscope (Fig. 1.7A).

- Latex gloves.
- Safety glasses.
- Fume hood.
- Rotator (Fig. 1.5C).

SOLUTIONS

Cacodylate Buffer – 0.5 mol/L

1. Add 107 g sodium cacodylate to a volumetric flask.
2. Dilute to 900 mL with distilled water.
3. Adjust pH to 7.3 with concentrated hydrochloric acid.
4. Dilute to 1,000 mL with distilled water.
5. Portion into labeled prescription or rectangular plastic bottles and store in a freezer. (Will keep indefinitely.)
6. Thaw and shake before use.

Glutaraldehyde – 2.5% in 0.1 mol/L Cacodylate Buffer (430 mosm)

1. Combine 6.4 mL 0.5 mol/L cacodylate buffer and 1 vial (10 mL) of 8% glutaraldehyde.
2. Dilute to 32 mL with distilled water.
3. Check pH – it should be 7.2–7.4.
4. Store in a 50-mL prescription bottle in the refrigerator.
5. Discard after 1 month.

or

1. Combine 10 mL 0.5 mol/L cacodylate buffer and 5 mL 25% glutaraldehyde.
2. Dilute to 50 mL with distilled water.
3. Check pH and store as above.

Glutaraldehyde – 1.5% in 0.1 mol/L Cacodylate Buffer

1. Combine 32 mL 0.5 mol/L cacodylate buffer and 30 mL 8% glutaraldehyde.
2. Dilute to 160 mL with distilled water.

or

1. Combine 20 mL 0.5 mol/L cacodylate buffer and 6 mL 25% glutaraldehyde.
2. Dilute to 100 mL with distilled water.

or

1. Combine 80 mL 0.2 mol/L cacodylate buffer and 30 mL 8% glutaraldehyde.
2. Dilute to 160 mL with distilled water.

or

1. Combine 50 mL 0.2 mol/L cacodylate buffer and 6 mL 25% glutaraldehyde.
2. Dilute to 100 mL with distilled water.

Formaldehyde–Glutaraldehyde (176 mosm) (McDowell & Trump, 1976)

1. Combine 1.16 g sodium biphosphate ($NaH_2PO_4 \cdot H_2O$) and 0.27 g sodium hydroxide (NaOH).
2. Dilute to 100 mL with distilled water.
3. Add 10 mL commercial formalin.
4. Add 2 mL 50% glutaraldehyde.
5. Mix by shaking.
6. Store at 4°C. (Stable for at least 3 months).

Cacodylate Buffer – 0.2 mol/L

1. Add 42.8 g sodium cacodylate to a volumetric flask.
2. Dilute to 900 mL with distilled water.
3. Adjust pH to 7.3 with concentrated hydrochloric acid.
4. Dilute to 1,000 mL with distilled water.
5. Portion into labeled prescription or rectangular plastic bottles and store in the freezer.
6. Thaw and shake before use.
7. For 0.1 mol/L solution, dilute 1:1 with distilled water.

Osmium Tetroxide – 2% in Distilled Water

1. Wash exterior of vial of osmium tetroxide crystals thoroughly with detergent and rinse well in distilled water to remove dirt and grease.
2. Put vial into a clean, lint-free amber glass bottle.
3. After replacing the lid, shake the bottle sharply to break the vial.
4. Open the bottle in a fume hood (with extractor fan running) and add 50 mL of distilled water.
5. Restopper tightly, wrap with "Parafilm," and when the osmium tetroxide has dissolved, store in the refrigerator.
6. Before use, filter into a clean amber bottle to remove glass chips.
7. At time of use, dilute 1:1 with 0.2 mol/L buffer, = 1% osmium tetroxide in 0.1 mol/L cacodylate buffer.

Notes on Solutions

- Glutaraldehyde should be stored in a refrigerator.
- Osmium tetroxide is reduced by dust, organic matter, or light. It dissolves slowly and will take several hours at room temperature to dissolve completely. The process can be speeded up by ultrasonication in a water bath for 15 minutes.
- Osmium tetroxide vapor is very hazardous to eyes and mucous membranes. It is advisable to work in a fume hood and wear protective glasses when handling this chemical.
- Discarded 2% osmium tetroxide may be neutralized by adding it to twice its volume of corn oil (Cooper, 1980).
- The following mixture should be kept in the laboratory at all times in case of a

Figure 1.8. Light micrograph of a 1 μm-section of liver showing damage done by permitting tissue to dry prior to fixation. Tissue shows fractures (f) and uneven staining and fixation.

spill: 100 g of kitty litter + 50 mL of corn oil. Osmium tetroxide (2%) is neutralized by six times its volume of this mixture, which may be stored in a sealed plastic bag.

- Almost all reagents used in the preparation of material for EM are toxic and should be handled carefully using a fume hood, latex gloves, and gas-tight goggles. Their disposal should follow safety guidelines (Barber & Clayton, 1985; Hayat, 1989, pp. 4–6).

- The pH of fixatives should be checked immediately before use (pH paper is adequate). Contaminated distilled water or a bad batch of glutaraldehyde can yield disastrous results. If unbuffered glutaraldehyde has a pH of less than 3.5–4.0, it should be discarded.

METHOD

1. Check the fixative pH either with a pH meter or with short-range pH paper. It should be pH 7.2–7.4.

2. Using iris forceps or flat toothpicks, gently place tissue in a pool of fixative (2.5% glutaraldehyde in 0.1 mol/L sodium cacodylate buffer) on a clean piece of dental wax.

3. Using a Pasteur pipette, quickly flood the tissue with more fixative (Fig. 1.5B). The tissue must not dry out even slightly (Fig. 1.8).

4. Place the dental wax and tissue on the stage of a dissecting microscope (Fig. 1.7A).
5. Trim the tissue, using a slicing motion (to prevent crushing damage), to 0.5 mm^3 (or 0.5-mm slices) with a double-edged razor blade that has been broken in half (break the blade in half before unwrapping it).
6. Gently support the tissue in a drop of glutaraldehyde between the tips of iris forceps or on a flat toothpick and place it in a 3.8-mL vial with 2.5% buffered glutaraldehyde. Each vial should contain a small pencil-written label (such as Lipshaw No. P-50) identifying the tissue and should be at least half filled with glutaraldehyde (Fig. 1.7A). The volume of fixative should be 10 to 20 times greater than the volume of tissue to be fixed. Details of each specimen should be recorded in a log book. See sample page in Appendix.
7. Fix the tissue in a refrigerator for 1 to 2 hours (overnight, if necessary).
8. After fixation, remove the vial of tissue from the refrigerator and pour the tissue and glutaraldehyde into a small weighing boat (Fig. 1.5B).
9. Transfer the tissue in a drop of glutaraldehyde (as in step 2) to a piece of dental wax (Fig. 1.5B).
10. With a clean, sharp razor blade and using a slicing (rather than crushing) motion, retrim the tissue (if necessary) to 0.5 mm^3. This should be done using a dissecting microscope to allow the identification and inclusion of specific areas of interest (e.g., kidney glomeruli).
11. Transfer the tissue to a vial containing 0.1 mol/L cacodylate buffer and place on a rotator (Fig. 1.5C).
12. After 10 minutes, pipette off the buffer and replace with fresh buffer.
13. After 5 minutes, pipette off this buffer and replace with 1% osmium tetroxide in 0.1 mol/L cacodylate buffer. Leave for 30 minutes to 1 hour on the rotator (osmium tetroxide acts not only as a secondary fixative but also as an electron-dense stain, rendering the tissue more visible to the naked eye).
14. Carry out the procedures of dehydration, infiltration with resin, and embedding (see Chapter 2). If dehydration and infiltration cannot be completed within a working day, tissues may be stored in glutaraldehyde in a refrigerator or rinsed in buffer, postfixed in osmium tetroxide, and stored in buffer.

Note on Method

- If fixation is carried out by perfusion, a prerinse with electrolyte solution is recommended, as fixed blood cells will block small vessels and prevent successful perfusion. Air bubbles in the perfusion apparatus must be avoided, as must excessive perfusion pressure. When perfusing a rat, a pressure of 15 960 PA (120 mm Hg) should be maintained for 10 minutes. If the perfusate is gravity fed, the fluid level should be 152 cm (61 in.) above the animal. Hayat (1989, pp. 4–6) has an excellent section on fixation by perfusion.

REFERENCES

Barber, V. C., & Clayton, D. L. (1985). *Electron Microscopy Safety Handbook*. San Francisco: San Francisco Press, pp. 1–33.

Barrnett, R. J., Perney, D. P., & Hagström, P. E. (1964). Additional new aldehyde fixatives for histochemistry and electron microscopy. *J. Histochem. Cytochem.* **12**: 36.

Bullock, G. R. (1984). The current status of fixation for electron microscopy: a review. *J. Micros.* **133**: 4.

Cooper, K. (1980). Neutralization of osmium tetroxide in case of accidental spillage and for disposal. *Bull. Microsc. Soc. Can.* **8**: 24.

Ghadially, F. N. (1980). *Diagnostic Electron Microscopy of Tumours.* London: Butterworth, p. 6.

Ghadially, F. N. (1985). *Diagnostic Electron Microscopy of Tumours*, 2nd ed. London: Butterworth, pp. 9, 10.

Gil, J., & Weibel, E. R. (1968). The role of buffers in lung fixation with glutaraldehyde and osmium tetroxide. *J. Ultrastruct. Res.* **25**: 331.

Hayat, M. A. (1986). *Basic Techniques for Transmission Electron Microscopy.* San Diego: Academic Press.

Hayat, M. A. (1989). *Principles and Techniques of Electron Microscopy: Biological Applications,* 3rd ed. Boca Raton: CRC Press.

Karnovsky, M. J. (1971). Use of ferrocyanide reduced osmium in electron microscopy. *Proc. 14th Ann. Meet. Am. Soc. Cell Biol.* p. 146a.

Lewison, D. (1989). Application of the ferrocyanide reduced osmium method for mineralizing cartilage: further evidence for the enhancement of intracellular glucogen and visualization of matrix components. *Histochem. J.* **21**: 259–270.

Luft, J. H., & Wood, R. L. (1963). The extraction of tissue protein during and after fixation with osmium tetroxide in various buffer systems. *J. Cell Biol.* **19**: 46A.

McDowell, E. M., & Trump, B. F. (1976). Histologic fixatives suitable for diagnostic light and electron microscopy. *Arch. Pathol. Lab. Med.* **100**: 405–414.

Sabatini, D. D., Bensch, K., & Barrnett, R. J. (1962). New means of fixation for electron microscopy and histochemistry. *Anat. Rec.* **142**: 274.

Sabatini, D. D., Bensch, K., & Barrnett, R. J. (1963). Cytochemistry and electron microscopy: the preservation of cellular structure and enzymatic activity by aldehyde fixation. *J. Cell. Biol.* **17**: 19.

Sabatini, D. D., Miller, F., & Barrnett, R. J. (1964). Aldehyde fixation for morphological and enzyme histochemical studies with the electron microscope. *J. Histochem. Cytochem.* **12**: 57.

Weakley, B. S. (1981). *A Beginner's Handbook in Biological Transmission Electron Microscopy.* 2nd ed. Edinburgh: Churchill Livingstone.

RECOMMENDED READING FOR CHAPTERS 1, 2, AND 3

Bullock, G. R. (1984). The current status of fixation for electron microscopy: a review. *J. Micros.* **133**: 1–15.

Hayat, M. A. (1986). *Basic Techniques for Transmission Electron Microscopy.* San Diego: Academic Press.

Hayat, M. A. (1989). *Principles and Techniques of Electron Microscopy: Biological Applications.* 3rd ed. Boca Raton: CRC Press.

Weakley, B. S. (1981). *A Beginner's Handbook in Biological Electron Microscopy.* 2nd ed. Edinburgh: Churchill Livingstone.

2

• *Dehydration and Embedding*

INTRODUCTION

Following fixation, water must be removed from the tissue and replaced with a medium that will withstand the stress of cutting. Dehydration may be accomplished using either alcohol or acetone, and there are many support media (acrylics, polyesters, and epoxy resins) on the market. The methods outlined are for epoxy resins: Spurr's resin (Spurr, 1969), Epon (Kushida, 1959), and Epon-Araldite (Weakley, 1981).

It has been my experience that *Spurr's resin*, being the least viscous, is easiest to handle. Tissue penetration is good, and block hardness can be varied by increasing (softer) or decreasing (harder) the flexibilizer (DER). It is less hygroscopic than Epon or Araldite and cuts very easily. Spurr's resin is stable under the electron beam, but image contrast is lower than with Epon or Epon-Araldite and the toxicity of ERL 4206 is greater than any component in the other two resins.

Epon penetrates well; block hardness can be varied; and beam stability and image contrast are good. *Epon-Araldite*, although more viscous and therefore less easy to handle, penetrates well, cuts better than Epon or Araldite alone, and has good electron-beam stability and image contrast.

Specimens are dehydrated by immersion in a graded series of alcohol or acetone, using short periods to minimize shrinkage and other tissue damage. En bloc staining with heavy metals (to increase tissue contrast) can be done during dehydration and is recommended, especially if Spurr's resin is used. Several methods can be found in the literature, one of which will be given later (Kushida & Fujita, 1966). Dehydration and infiltration are carried out at room temperature using a rotator, if an automatic processor is not available. Specimen vials should be filled to ensure that the tissue is covered with fluid at all times while rotating.

During fixation, dehydration, and infiltration, fluids are removed from the vials of tissue with Pasteur pipettes. Care must be taken to avoid accidentally drawing up the tiny pieces of tissue and thus discarding them or transferring them from one vial to another. Both the inside and outside of the pipette should be checked carefully at each step. Fluids must be replaced quickly because tissue must not be allowed to dry (Fig. 2.1).

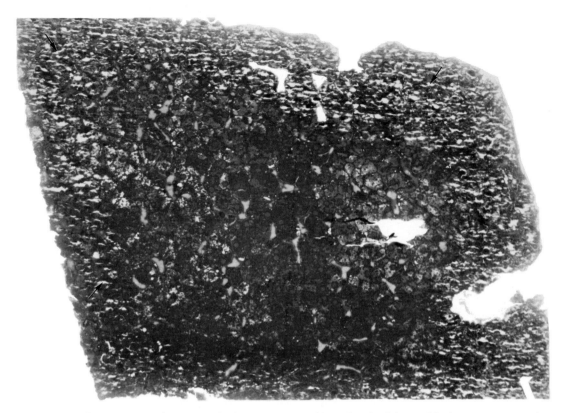

Figure 2.1. Light micrograph of a 1 μm-section of liver that dried during dehydration. Consequently, periphery of block has not infiltrated with resin and shows numerous fractures (*arrows*).

Reagent-grade solvents must always be used. Acetone and/or alcohol, for final dehydration, may be stored over a molecular sieve to ensure that it is completely free of water. Any water trapped in the tissue causes incomplete infiltration of the support medium, resulting in blocks that cut poorly, stain unevenly, and contain air pockets.

Alcohol (usually ethanol) is the solvent most frequently used for dehydration. Some resins (e.g., Epon, Araldite) are not readily miscible with alcohol, and a transitional solvent is recommended. Propylene oxide (epoxypropane) is used for this purpose in many laboratories. However, it is very toxic, has a low flash point, and is not recommended; reagent-grade acetone should be used instead. In cases where a transitional solvent is unnecessary (e.g., Spurr's resin), acetone is still recommended following alcohol dehydration since it can be purchased in a very pure form, thus decreasing the chances of the moisture contamination that can occur with alcohol.

Since epoxy resins have been shown to be carcinogenic (Causton et al., 1981), they should be handled with care using latex gloves. Alcohol facilitates skin penetration of the resins and exposed skin should be washed with soap and water. Resin spills should be cleaned immediately, with alcohol or acetone, keeping hands protected. Cured resin can be dissolved with Plastisolve or sodium ethoxide (saturated NaOH in absolute ethanol) but it is best to avoid spills, if possible.

CHEMICALS AND EQUIPMENT

- Pasteur pipettes – 145 mm (5 in.), rubber bulbs.
- Plastic wash bottles – 250 mL, four or five (Fig. 2.2A).
- Amber glass bottles – 200 mL (Fig. 2.2A).
- Prescription bottles – 150 mL (Fig. 2.2A).
- Weighing boats – 35 mm (1.5 in.) (Fig. 2.4).
- Iris forceps – stainless steel (Fig. 2.4A).
- Microspatula (Fig. 2.4A).
- Paper labels in perforated sheets (e.g., Lipshaw No. P-50) 7 × 20 mm (9/32 in. × 13/16 in.) (Fig. 2.4A).
- BEEM (Better Equipment for Electron Microscopy) capsules – 5.2-mm diameter, or desired size (Fig. 2.2B).
- BEEM capsule holders or Falcon 3070 flat-bottom microtiter plates (Fig. 2.2B).
- Transfer pipettes, polyethylene – 5 mL (Fig. 2.2B).
- Plastic bottles, rectangular – 125 mL (Fig. 2.2A).
- Syringe – 20 mL with 16-gauge blunt needle (Fig. 2.4A).
- Graduated cylinders – 5, 10, 25, 50, 100, 250 mL.
- Balance, single-pan, top-loading accurate to 2 decimal places (Fig. 2.3).
- Acetone – certified 99.5 mol % pure.
- Acetone – technical grade.
- Ethanol – absolute.
- Molecular sieve – type 4A 1.5 mm (1/16 in.), (8–12 mesh).
- BEEM block storage case or small, white sliding-form pill boxes (Fig. 2.2C).
- Storage cabinet for above (Fig. 2.2D).
- Labels, small self-adhesive type.
- Dissecting needle or straight pin to orientate tissue if necessary (Fig. 2.5B).
- Plastisolve.
- Oven, standard convection type with a range from 50°C to 200°C.
- Latex gloves.

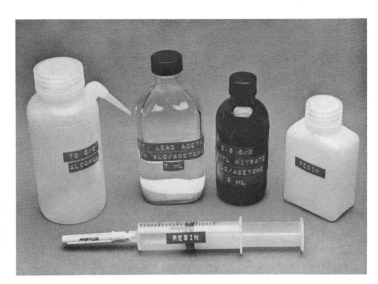

Figure 2.2. *Materials recommended for dehydration and embedding.*

A. Bottles for ethanol series, en bloc staining solutions, and hypodermic syringe of resin.

(Fig 2.2 B, C, and D on p. 20)

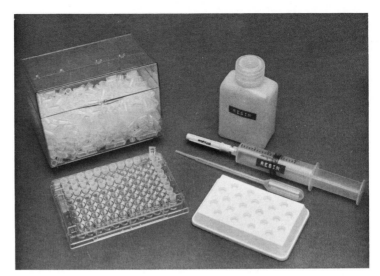

B. BEEM capsules, embedding trays, transfer pipette, and syringe with blunt needle.

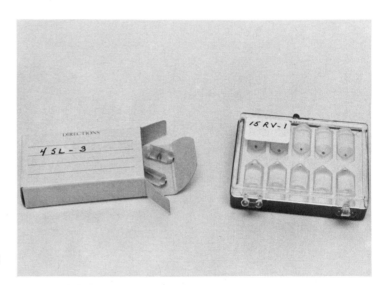

C. BEEM and pill boxes for storing polymerized blocks.

D. Storage cabinet.

- Pen, with non-water-soluble ink and pencil.
- Uranyl nitrate – $UO_2(NO_3)_2 \cdot 6H_2O$.
- Lead acetate – $Pb(C_2H_3O_2)_2 \cdot 3H_2O$.
- Laboratory film (Parafilm) – 102 mm × 3,810 cm (4 in. × 125 ft) roll.

Epon-Araldite

- 812 resin, epoxy monomer – 250 mL.
- Araldite 502 resin, epoxy monomer – 250-mL bottles.
- DDSA (dodecenyl succinic anhydride), hardener – 500 mL.
- DMP-30 (2, 4, 6-tridimethylamino methylphenol), accelerator – 50 mL (or smaller if possible)

or

- BDMA (benzyldimethylamine), accelerator – 50 mL (or smaller, if possible).

Epon

- 812 resin.
- DDSA.
- MNA (methyl nadic anhydride), hardener – 500 mL.
- BDMA or DMP-30.

Spurr's Resin

- ERL 4206 (VCD) resin (epoxy monomer, vinylcyclohexene dioxide) – 250 mL.
- DER 736 (diglycidyl ether or propylene glycol), flexibilizer – 250 mL.
- NSA (nonenyl succinic anhydride), hardener – 500 mL.
- S-1 (DMAE) (dimethylaminoethanol), accelerator – 50 mL (or smaller if possible).

SOLUTIONS – EN BLOC STAINING (KUSHIDA & FUJITA, 1966)

Uranyl Nitrate Solution

1. Prepare 0.5 g uranyl nitrate made up to 100 mL with alcohol-acetone (1:1).
2. Store in an amber glass bottle. (If protected from light and moisture, reagent will keep at least a year.)

Lead Acetate Solution

1. Pour lead acetate to a depth of approximately 1 cm in a 150-mL prescription bottle.
2. Fill bottle with absolute ethanol-acetone (1:1) and shake vigorously.
3. Prepare solution 1 day before use to allow time for lead acetate to settle.
4. After two or three refills of ethanol-acetone, discard solution (and bottle) and prepare fresh solution.
5. Mix solutions in the proportion of uranyl nitrate: lead acetate—3:7 at time of use.

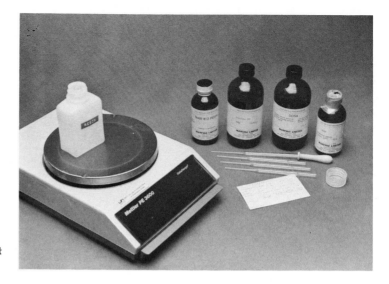

Figure 2.3. Chemicals and equipment required to prepare resin.

EMBEDDING MEDIA

Spurr's Resin

VCD (ERL 4206)	10.0 g
DER 736	6.0 g
NSA	26.0 g
DMAE (S-1)	0.4 g

NOTE: This is a standard mixture. For a hard mixture, use 4 g DER 736; for a soft mixture, use 8 g DER 736.

Epon

Epon 812	24.0 g
DDSA	12.0 g
MNA	14.0 g
DMP-30 (or BDMA, less viscous)	0.6 g
(Glauert, 1987)	

NOTE: The hardness of the blocks may be varied by changing the relative amounts of the two hardeners. Increasing MNA yields a harder block.

Epon-Araldite

Epon 812	12.5 g
Araldite 502	10.0 g
DDSA	30.0 g
BDMA (or DMP-30)	0.85 g

NOTE: This yields a firm block with excellent cutting qualities. Increasing Epon or decreasing Araldite will yield a softer block.

1. Prepare resins by adding the components gravimetrically, in the above order, to a plastic bottle (Fig. 2.3). Make certain that the lid is on securely and shake the bottle *well* before adding accelerator (BDMA), then again after accelerator has been added.
2. Weigh chemicals accurately (use a pipette to add the last few milliliters) and *mix thoroughly.*

NOTE: Resins are best prepared the same day they are to be used but if that is impossible, they may be drawn into cleaned plastic syringes and frozen for up to 1 week. Always bring the resins to room temperature before use to avoid contamination by condensation. (Resin may be thawed and refrozen, but should be discarded if there is any likelihood of water contamination.)

Notes on Solutions

- Lead and uranium are toxic chemicals. Avoid skin contact and inhalation. Dispose of solutions according to safety regulations. DO NOT pour solutions the sink!
- During preparation and use of resins, exposure to atmospheric moisture must be minimized. Avoid mixing in a wide-mouthed container or with a wooden spatula.
- Resin components are toxic and some are carcinogenic. They must be handled with care. Avoid skin contact with the various components, and avoid breathing their vapors.
- Resin may be disposed of safely by discarding it in a closed waste container in the fume hood. When nearly full, the container may be placed in a 70°C oven to polymerize the resin.
- Barrels and plungers of hypodermic syringes are frequently contaminated with grease. Plungers should be wiped clean and the barrels rinsed with acetone to avoid contaminating resin. They must be completely dry before use.

METHOD

All dehydration and infiltration steps should be done on a rotator.

1. Pipette the osmium tetroxide from each of the vials of tissue, replacing it with 0.1 mol/L cacodylate buffer for 5–10 minutes.
2. In this manner, dehydrate through 70%, 95% and absolute ethanol. Give the tissue two or three changes in each grade of ethanol, 5 minutes in each change.
3. Give the tissue two changes of absolute ethanol–acetone (1:1) over 10 minutes.
4. During this time, prepare the en bloc stain in a ratio of 3:7, uranyl nitrate:saturated lead acetate (Fig. 2.2A).
5. Replace the final absolute ethanol–acetone with the en bloc stain and leave for 1 hour.
6. During this time, prepare resin (Fig. 2.3), or bring frozen resin to room temperature.
7. Give the tissue two changes of 1:1 absolute ethanol–acetone over 10 minutes.
8. Give the tissue two changes of acetone over 10 minutes.
9. Replace acetone with 30% resin in acetone for 30 minutes.
10. Replace with 50% resin in acetone for 45 minutes.
11. Change to 70–80% resin in acetone for 1 hour (or overnight if desired).

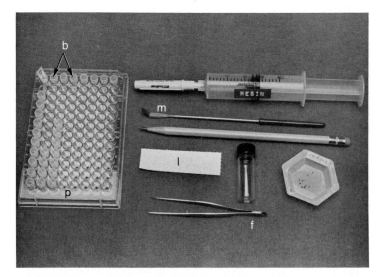

Figure 2.4. *Embedding.*
  A. Preparing to embed. Microtiter plate (p), BEEM capsules (b), iris forceps (f), labels (l), bent microspatula (m).

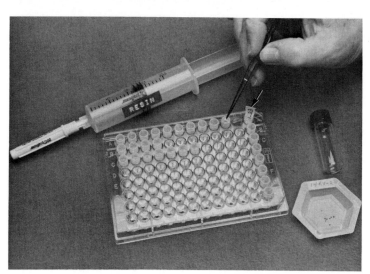

  B. Centering the tissue. Label (*arrow*) clearly idetifies each group of tissue.

12. After pipetting off this resin mixture, pour pure resin into the specimen vial and rotate for 1 hour (or overnight).
13. Remove the vial of tissue from the rotator and, while the tissue is still suspended in resin, pour it and the label into a small weighing boat (Fig. 2.4A). Should a piece of tissue stick to the bottom of the bottle, it may be lifted out gently using a microspatulla whose tip has been bent at a 45° angle (Fig. 2.4A).
14. Remove the lids from BEEM capsules, place the capsules in a holder, and fill them with fresh resin. [Dispensing resin from a 20-mL syringe fitted with a blunt 16-gauge needle and placing the BEEM capsules over a black mat will lessen the chances of overflow of the capsules (Fig. 2.4A). Heating the resin 2–3 minutes in a 70°C oven will make it less viscous and easier to handle. If the capsules are filled from the top, air bubbles can form at the base. This can be avoided by introducing the needle deep into the capsule and filling it from the bottom.]
15. Gently pick up a piece of tissue with iris forceps, blot it lightly on the rim of the

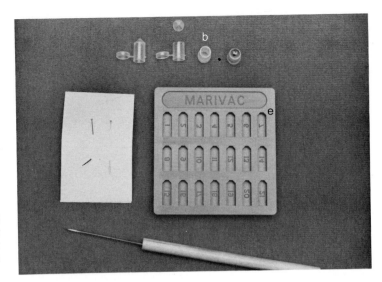

Figure 2.5. *Embedding.*
A. Flat embedding mold (e). BEEM capsules that have been prepared for use as flat embedding molds (b). Straight pins and a dissecting needle to aid in specimen orientation.

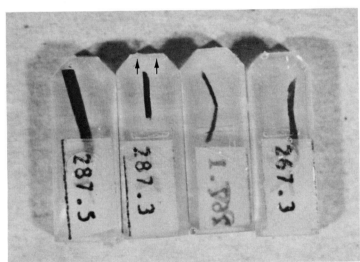

B. Carelessly embedded tissue. Excess plastic (*arrows*) will have to be trimmed away, resulting in wasted time and unnecessary exposure to epoxy dust.

weighing boat and center it at the top of a BEEM capsule (Fig. 2.4B). The tissue will sink to the bottom on its own, leaving behind any residual acetone. Recenter the tissue (if necessary) using a dissecting needle or pin.

16. Place embedded and labeled tissues in a 70°C oven and cure for 6 to 8 hours or overnight. The oven must be vented or used in a fume hood.
17. After removing the cured blocks from the oven, they may be labeled with a non-water-soluble pen. The blocks are now ready to be cut (Fig. 2.2C) or may be stored in properly labeled BEEM or sliding-type pill boxes (Fig. 2.2D).

Notes and Method

● Osmium tetroxide, dehydrants, heavy metal stains, and resin should be disposed of in properly labeled waste containers in accordance with locally prescribed safety standards. DO NOT pour them down the sink!

- Pasteur pipettes should be checked for adherent tissue after use in each specimen vial. Mix-ups of tissue from two different patients can have serious consequences if diagnoses are dependent upon EM findings.
- Since they have a larger bore than Pasteur pipettes, hypodermic syringes and plastic transfer pipettes are more convenient for dispensing pure resin.
- Tissues should *never be permitted to dry* (Fig. 2.1). When pipetting solutions, be sure to leave a residue to cover the tissue.
- Avoid skin contact with osmium tetroxide, heavy metal stains, and resin.
- If possible, set up the rotator beside or within a fume hood. Change solutions and embed with the fume hood running.
- It is essential that dehydrants be kept dry. If this is a problem, store them over a molecular sieve or silica gel.
- Do not use the same pipette in different solutions (e.g., the pipette used to dispense buffer should not be used to pipette osmium tetroxide).
- If exact orientation of the tissue is necessary, it must be embedded using a dissecting microscope. Use a dissecting needle or straight pin (Fig. 2.5A) to gently move the tissue to the correct position. Sometimes larger pieces of tissue are best handled this way (e.g., a cross section of muscle).
- In some cases, it is necessary to use flat embedding molds, but in most cases, embedding in the lid of a BEEM capsule (after removing the pyramidal end with a sharp razor blade) permits excellent orientation (Fig. 2.5A). Fig. 2.5B shows tissue that has been embedded carelessly.
- Small BEEM capsules (e.g., 5.2 mm), while requiring less resin, may be more difficult for some people to handle than the larger sizes. However, the tip of a BEEM capsule is $1 \, mm^2$ regardless of its diameter.
- Specimens embedded on edge, which later fall over during curing, may be cut out and reembedded as follows. Hold the block in a vise and, using a jeweler's saw, cut out the specimen using very straight cuts. This produces flat sides that are stable for reembedding. Wear a mask while sawing and, if possible, place a vacuum cleaner nozzle as close to the vise as possible. Avoid breathing resin dust.
- If very dense tissue is being processed (e.g., collagen, muscle, nerve), it is best to infiltrate with resin (see Steps 9–11) over a long period. In this case, use acetone-resin (3:1) for several hours, 1:1 overnight, and then pure resin for 4 to 6 hours, or until the next day.
- BEEM capsule holders (Fig. 2.2B) may be cleaned by soaking them in technical-grade acetone in a sealed container inside a fume hood.
- Carefully identify the embedded tissues. This can be done by placing tiny pencil-written labels around the inside of the BEEM capsule near the top. It is a very safe method but very tedious unless one uses BEEM capsules that are larger than 5.2 mm. With care, tissues may be identified by placing an empty BEEM capsule containing a small label with each group of tissues (Fig. 2.4B), or a record may be kept showing the exact location of each tissue in the capsule holder.

SUMMARIZED SCHEDULE OF FIXATION, DEHYDRATION, AND EMBEDDING

1. Check the pH of glutaraldehyde.
2. Trim tissue to 0.5-mm strips and fix in 2.5% glutaraldehyde 2 hours or overnight.
3. Retrim to $0.5 \, mm^3$.

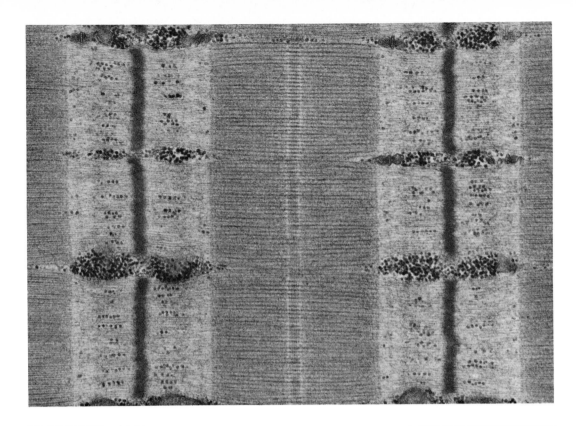

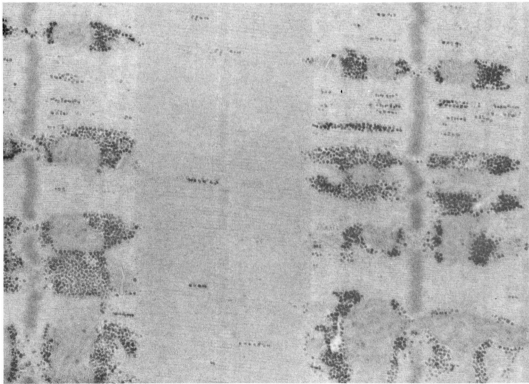

Figure 2.6. *Processing.*
    Top. Human muscle processed routinely. **Bottom.** Another portion of same tissue processed rapidly and cured in a microwave oven.

4. Wash in two changes of 0.1 mol/L buffer over 15 minutes.
5. Fix in 1% osmium tetroxide in 0.1 mol/L buffer, 30 minutes to 1 hour.
6. Place in 0.1 mol/L buffer for 5 to 10 minutes.
7. Using 70% ethanol, make two to three changes over 10 to 15 minutes.
8. Using 95% ethanol, make two to three changes over 10 to 15 minutes.
9. Using absolute ethanol, make two to three changes over 10 to 15 minutes.
10. Using absolute ethanol–acetone (1:1), make two changes over 10 minutes.
11. Place in uranyl nitrate–lead acetate (3:7) for 1 hour.
12. Using absolute ethanol–acetone (1:1), make two changes over 10 minutes.
13. Using acetone, make two changes over 10 minutes.
14. Infiltrate with 30% resin in acetone for 30 minutes.
15. Infiltrate with 50% resin in acetone for 45 minutes.
16. Infiltrate with 70–80% resin in acetone for 1 hour or overnight.
17. Infiltrate in pure resin for 1 to 3 hours or overnight.
18. Embed and label.
19. Cure in a 70°C oven for 6–8 hours or overnight.

RAPID SCHEDULE

1. Prefix 0.5 mm strips with buffered 2.5% glutaraldehyde for 10 minutes at room temperature.
2. Retrim to 0.5 mm^3 or less and fix a further 20–30 minutes.
3. Wash with suitable buffer for 5 minutes.
4. Postfix with buffered osmium tetroxide for 30 minutes.
5. Rinse with buffer 5 minutes.
6. Dehydrate in ethanol series (70%, 95%, and two changes of 100%), 5 minutes each change.
7. Rinse in absolute ethanol–acetone (1:1) for 10 minutes.
8. Rinse in acetone for 10 minutes.
9. Infiltrate with 30% resin in acetone for 15 minutes.
10. Infiltrate in 50% resin in acetone for 15 minutes.
11. Make two changes of pure resin for 15 minutes each.
12. Embed in fresh resin in silicone molds and polymerize for 20 minutes in a 640 watt microwave oven at 80% power or 1 hour at 100°C in a conventional oven. If Spurr's resin is used, it is recommended to en bloc stain in uranyl nitrate–lead acetate (3:7) for 1 hour between steps 7 and 8 as in the routine schedule. Figure 2.6 shows a comparison of muscle processed routinely (top) and rapidly with microwave embedding (bottom).

REFERENCES

Causton, B. E., Ashhurst, D. E., Butcher, R. G., Chapman, S. K., Thomson, D. J., & Webb, M. J. W. (1981). Resins: toxicity, hazards and safe handling. *Proc. Roy. Microsc. Soc.* **16**: 265–268.

Glauert, A. (1987). Accelerators for epoxy resins. *Proc. Roy. Microsc. Soc.* **16**: 264–265.

Kushida, H. (1959). On an epoxy resin embedding method for ultrathin sectioning. *Electron Microscopy.* **8**: 72–74.

Kushida, H., & Fujita, K. (1966). Simultaneous double staining. *Proc. 6th Inter. Cong. Electron Microsc.* **2**: 39.

Spurr, A. R. (1969). A low-viscosity epoxy resin embedding medium for electron microscopy. *J. Ultrastruct. Res.* **26**:31.

Weakley, B. S. (1981). *A Beginners Handbook in Biological Transmission Electron Microscopy*, 2nd ed. Edinburgh: Churchill Livingstone, p. 221.

3

• *Cutting*

SEMITHIN SECTIONS

Introduction

Cutting semithin sections prior to thin sections is useful for two reasons. It indicates whether or not fixation, dehydration, and infiltration have been carried out properly, and it helps to locate areas of interest (e.g., glomerulus, lesion, specific cell type). This avoids unnecessary thin sectioning of unsuitable blocks. Semithin sections of tissue processed for electron microscopy also provide excellent material for photomicrography (Fig. 3.1). Fixation is always superior to material fixed for conventional light microscopy and the sections (0.5–2 μm), provide better resolution than paraffin sections, which are usually about 5 μm.

Semithin sections are cut with a glass knife, which is fitted with a trough to hold distilled water, and are usually stained with alkaline toluidine blue (Trump, Smuckler, & Benditt, 1961). A silver staining method that works well on resin embedded tissue (Rosenquist, Slavin, & Bernick, 1971) is also included here [Fig. 3.1 (bottom)].

It is important to be familiar with all of the ultramicrotome controls. If a manual is not available, request one from the manufacturer.

Chemicals and Equipment

Knife-Preparation Equipment

- Knife maker (Fig. 3.2A).
- Glass strips — 25 mm × 400 mm × 6.4 mm (1 in. × $1\frac{5}{8}$ in. × $\frac{1}{4}$ in.) (Fig. 3.2B).
- Towels, lint-free.
- Liquid detergent.
- Paintbrush — 25 mm (1 in.).
- Dissecting microscope.
- Plastic tape — 10 × 33 m ($\frac{3}{8}$ in. × 100 ft) (Fig. 3.3).
- Razor blades, single-edged (Fig. 3.3).
- Nail polish, clear (Fig. 3.3).
- Blunt-end forceps — 125 mm (5 in.) (Fig. 3.3).
- Perspex box to hold glass knives (Fig. 3.5).

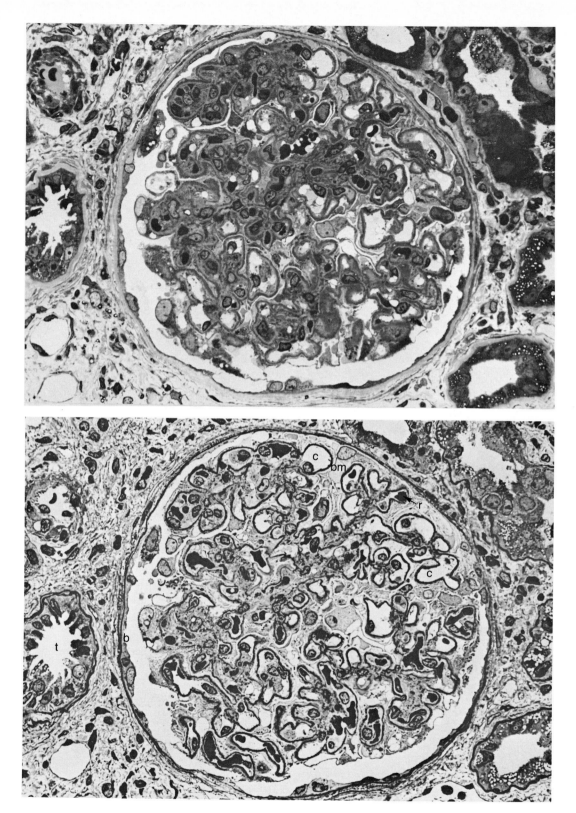

Figure 3.1. *Photomicrographs of semithin sections.*
   Top. 1 μm section of human kidney, embedded in Epon-Araldite and stained with 1% toluidine blue.
Bottom. 1 μm section of same kidney, embedded in Epon-Araldite and stained with silver nitrate. Visible
are Bowman's capsule (b), external lamina (bm), capillary loop (c), erythrocytes (r), tubule (t).

Slide-Coating Equipment

- Chrome alum (chromic potassium sulfate).
- Gelatin.
- Tissue, lint-free (Fig. 3.7).
- Microscope slides – 25 mm × 75 mm (1 in. × 3 in.), high quality, precleaned.
- Hydrochloric acid, concentrated.
- Ethanol, absolute.

Semithin-Section Cutting Equipment

- Ultramicrotome–Reichert Ultracut is used by the author.
- Razor blades, double-edged, halved, degreased with ethanol.
- Iris forceps.
- Beakers – 50, 100, 250, 400 mL.
- Hot plate (Fig. 3.8).
- Pasteur pipettes – 145 mm (5 in.) with rubber bulbs.
- Platinum wire – 8 mil × 0.6 m (2 ft) roll (Fig. 3.13).
- Alcohol burner.
- Dissecting needles.
- Applicator sticks.
- Coarse hair, taped to an applicator stick (Fig. 3.13).
- Toothpicks, flat.
- Prescription bottles – 50, 150 mL (Fig. 1.5A).
- Ethanol, absolute.
- Hypodermic syringe – 20 mL.
- Filter – 0.22-μm pore size (e.g., Gelman 4418) (Fig. 3.8).
- Wash bottles, Nalgene – 250 mL.

Semithin-Section Staining Chemicals and Equipment

- Toluidine blue – 50 g.
- Borax (sodium borate) – 50 g.
- Filter funnel – 50 mm (2 in.), short form.
- Filter paper – Whatman No. 1, 4.25 and 12.5 cm.
- Erlenmeyer flask – 1,000 mL.
- Coplin jars – 12, 50 mL.
- Silver nitrate – 50 g.
- Gold chloride – 10 g.
- Hydroquinone – 50 g.
- Gelatin powder – 100 g.
- Sodium thiosulfate – 100 g.
- Oxalic acid – 50 g.
- Tuberculin syringe – 1 mL.
- Hypodermic needles – 22 gauge (0.7 mm), 1.5 in. (40 mm).
- Hypodermic syringe – 20 mL.
- Filter – 0.45 μm (e.g., Gelman 4217) (Fig. 3.8).

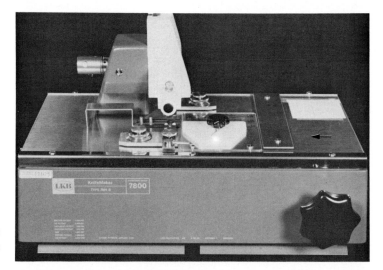

Figure 3.2. *Making glass knives.*
A. LKB knife maker. Arrow points to a dot used as a guide when breaking strips in half.

B. Three stages of knife making: half strip, square, and knives.

Making Glass Knives Using an LKB Knife Maker – Method

1. Read the instructions for the knife maker. They should be kept on a shelf at the base of the instrument. If there are no instructions, write to the manufacturer for a set. Familiarize yourself with the instrument before attempting to make knives. Use the type of glass recommended by the manufacturer.
2. Break long strips of glass in half (manufacturer's score facing down) using the guide (●) on the knife maker [Fig. 3.2A (arrow)].
3. Wash the glass in warm water and detergent.
4. Rinse, then dry with a lint-free towel.
5. Store in a dust-free box near the knife maker.
6. Set the knife guide along the 90° mark (for 45° knives).
7. Put the forklift in place.
8. Set the scoring guide to

C. Correctly scored square showing the score line ending 1–2 mm (*arrow*) from the tip of the glass.

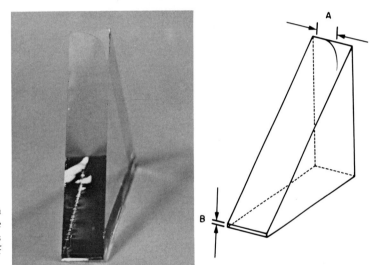

D. Correctly broken knife with straight cutting edge and base opposite (B) of less than 0.5 mm., (A) indicates sharpest and most score-free section of knife.

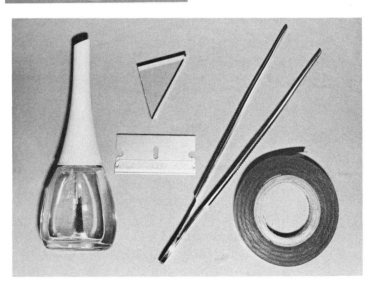

Figure 3.3. Materials for making troughs on glass knives.

9. Place a strip of clean glass (manufacturer's score down) in the knife maker, sliding it along the guide to the metal stop.
10. Ensure that the glass strip is tight against the guide, but do not apply downward pressure on it.
11. Lower the clamping head firmly, pull the scoring wheel forward, and break the glass. This will produce a 25-mm square (Fig. 3.2B).
12. Remove the remaining glass strip.
13. Turn the square $\frac{1}{4}$ turn clockwise and clamp in position.
14. Set the scoring guide to 25 $\Diamond$.
15. Lower the clamping head firmly, then pull the scoring wheel forward. Figure 3.2C shows a correctly scored square.
16. Position damping device, if knife maker has one.
17. Ensuring that the forklift is under the square, break the glass.
18. Raise the clamping head with the left hand while supporting it with the right.
19. Release the rear glass holder and move the fork to the left, making sure that the two knives do not touch each other.
20. Remove the knives from the fork, touching only the outside corners.
21. Inspect the knives (with the cutting edge up) under a dissecting microscope. Direct a light source along this edge and focus. The left half of the knife should show a straight, bright line with some feathered streaks appearing only to the right of center. The edge should be straight or slightly covex (Fig. 3.2D).

Notes on Method

- Keep the knife maker free of chips of glass at all times. A 25-mm (1 in.) paintbrush is recommended for chip removal.
- Wash hands before making glass knives (or wear lint-free gloves).
- Waste glass and unsatisfactory knives should be discarded in a metal tin or bucket, not in the wastepaper basket.

Making Troughs (Boats) on Glass Knives – Method

The materials necessary to make boats on glass knives are shown in Fig. 3.3.

1. Be very certain that hands are clean before making boats.
2. Check that the end of the plastic tape [which should be no wider than 10 mm ($\frac{3}{8}$ in.)] is cut square.
3. Apply the tape to the left-hand side of the knife as shown in Fig. 3.4 (top left), being sure neither fingers nor tape touch the cutting edge. Handle the knife by its base.
4. Gently (do not pull too taut) wrap the tape around the knife as shown in Fig. 3.4 (top right), finishing along the right-hand edge [Fig. 3.4 (bottom left)].
5. Slit the tape from bottom to top as shown in Fig. 3.4 (bottom right), being careful not to touch the cutting edge of the knife with the razor blade.
6. Press the tape firmly along the sides of the knife between thumb and index finger. If the tape has been applied properly, the new end should be square, ready to apply to the next knife; if not, straighten the end.

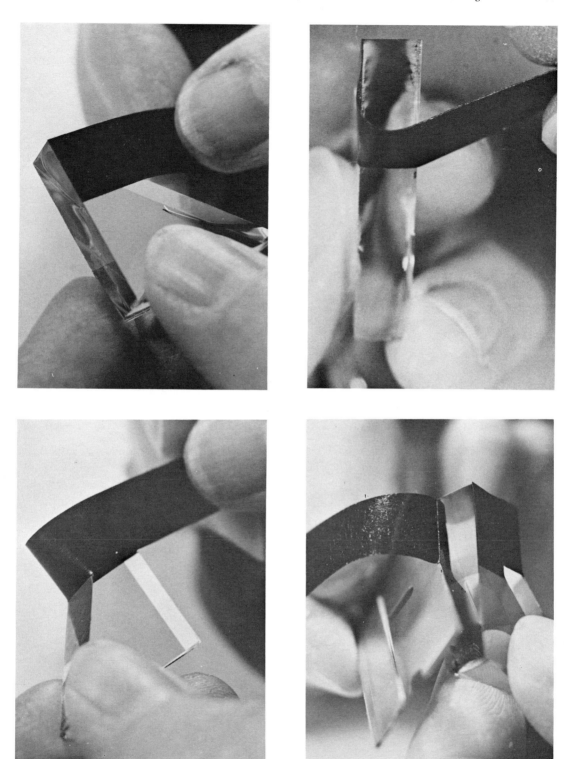

Figure 3.4. *Making a trough on a glass knife.*
Top left. Applying tape to left side of knife. **Top right.** Wrapping tape gently around knife. **Bottom left.** Securing tape along right side of the knife. **Bottom right.** Cutting away excess tape from bottom to top.

Figure 3.5. *Sealing trough with nail polish.*
Left. Applying nail polish to the base "B" of the tape. **Right.** Sealing along sides "A" and "C" (not visible).

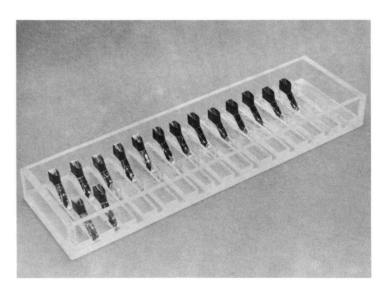

Figure 3.6. Perspex storage box.

7. As shown in Figs. 3.5 (left) and 3.5 (right), apply clear nail polish to area "B" and along sides at the base of the tape. The boat is most likely to leak at the area "B", so apply the nail polish liberally to that area (Fig. 3.5 left).

8. Place the knives in a protective storage box with a lid such as the Perspex one shown in Fig. 3.6.

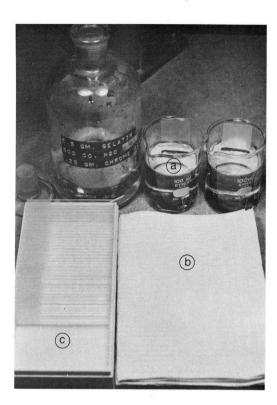

Figure 3.7. Coating slides with chrome gelatin. They are shown dipped in chrome gelatin (a), drained on paper towel (b), stored in slide box (c), and dried in oven.

Chromic Gelatin Coating of Slides (Rogers, 1973)

Chrome Gelatin Solution

1. Dissolve 2.5 g of gelatin in 500 mL of distilled water in an Erlenmeyer flask under a stream of hot running water or using a heated magnetic stirrer. (NOTE: The gelatin *must* be dissolved before adding the chrome alum in the next step.)
2. Add 0.25 g chrome alum (chromic potassium sulfate).
3. Filter before use. (Solution will keep 48 hours, if refrigerated.)

Method

1. Soak slides overnight or longer in absolute ethanol containing approximately 0.3% concentrated hydrochloric acid.
2. Towel dry slides, one at a time. They must be dirt and streak free. Dip slides in chrome gelatin (Fig. 3.7). Drain as shown in Fig. 3.14A, place them in a slide box, and dry in a 60°C oven for at least 30 minutes.

Semithin-Section Cutting

Toluidine Blue Solution

1. Combine 1 g toluidine blue and 1 g borax (sodium borate) in 100 mL distilled water.
2. Since the stain improves with age, make 1 liter and portion it into 6-oz (150-mL) prescription bottles.

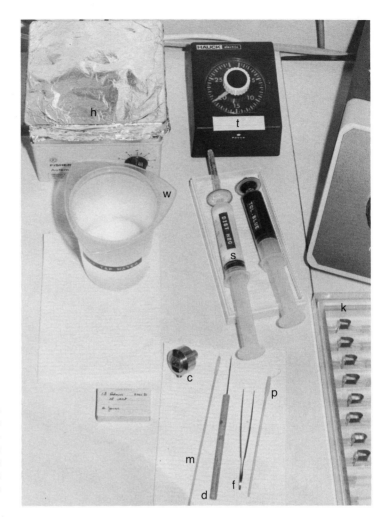

Figure 3.8. Convenient cutting area showing hot plate (h), timer (t), hypodermic syringes fitted with filters (s), beaker of hot tap water (w), forceps (f), platinum loop (p), dissecting needle (d), and hair (m). Block of tissue in a chuck (c) and box of knives (k) are also shown.

3. Label and date the stain.
4. Store at room temperature.

Method

Figure 3.8 illustrates a convenient cutting area.

1. Set the hot plate to 80°C. (A hot plate can be precalibrated and the various temperatures marked on the dial.)
2. Remove a block from a BEEM capsule by slitting the capsule in two places with a single-edged razor blade (Fig. 3.9A) and peeling the capsule away (Fig. 3.9B). [Keep fingers clear of the razor blade or use a BEEM capsule press (Fig. 3.9C).]
3. Set the block into a microtome chuck and remove excess resin from around the tissue. Trim as close as possible to the tissue laterally and try to keep top and bottom edges parallel (Fig. 3.10A). Use a dissecting microscope or the facilities on the ultramicrotome for this purpose and trim with a razor blade that has been broken in half and cleaned thoroughly with ethanol. Figure 3.10B shows a trimming block constructed of 0.5 in. brass.

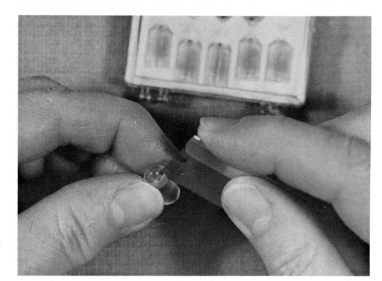

Figure 3.9. *Removing BEEM capsules from plastic blocks.*
    A. Slitting capsule with razor blade.

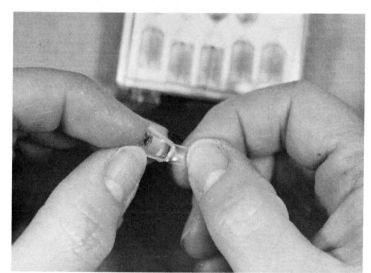

B. Peeling capsule from block.

C. BEEM capsule press.

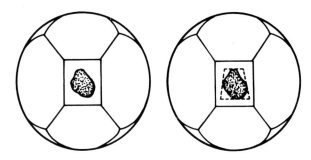

Figure 3.10. *Trimming a specimen block.*

A. Trim away excess plastic as indicated by the dotted line.

B. Brass trimming block.

4. Set the microscope magnification to 10 ×.

5. Put the specimen holder in position on the ultramicrotome.

6. Put a glass knife in position and adjust the knife angle (usually 2° to 6°).

7. With an old knife or the far right-hand side of a new knife, trim away excess resin to expose the tissue. This can be done either with a dry knife or with a boat in place filled with distilled water. This trimming should be done gently, or the tissue will be damaged. Remove only 1 or 2 μm at a time.

8. Unwanted sections can be removed from the boat by raising the meniscus and carefully dragging a dissecting needle or a piece of cellulose tape across the surface, parallel to the knife edge, from front to back, pulling the sections over the boat (Fig. 3.11A). Be very careful not to touch the knife edge. Remove the sections from the dissecting needle with a cellulose wipe moistened with ethanol.

9. When the tissue has been exposed, retrim the block as described in step 3, and reposition it in the microtome. NOTE: If one is searching for a specific area (such as a kidney glomerulus), the tissue should be exposed with a wet knife and rough sections checked every 30–40 μm until the desired area is reached. (Alternatively, a mesa may be used. See note "l" in following section and Fig. 3.11B.)

10. Take a new knife or the unused portion at the left-hand side of the trimming knife.

11. Fill the trough with filtered distilled water.

12. While viewing through the microscope, adjust the meniscus until the surface appears silver. Meniscus adjustment is best learned by trial and error but if the

Figure 3.11.
A (top). Excess sections eliminated with a dissecting needle (as shown) or with a strip of cellulose tape.

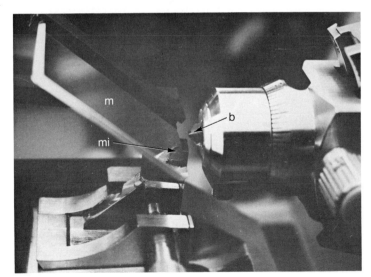

B (middle) and **C** (bottom). Mesa used to facilitate searching for discrete structures such as renal glomeruli. Shown are magnetic stand (s), mirror (m), articulated arm (a). Block (b), and mirror image of block face (mi).

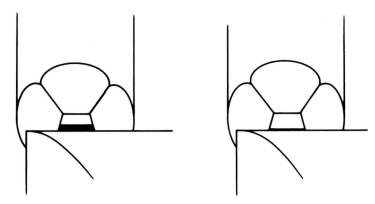

Figure 3.12. *Approximating knife and tissue block.*
Left. Upper portion of block face appears silver; lower portion appears black. **Right.** Knife and block are brought together until black portion is barely visible.

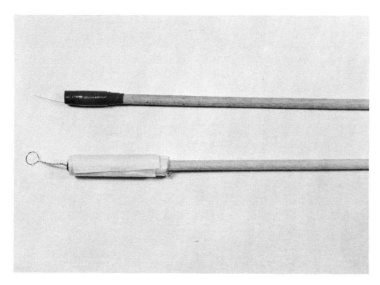

Figure 3.13. Top. Coarse hair to tease sections. **Bottom.** Platinum wire loop to pick up sections

meniscus is too low, the knife edge will be dry and the sections will compress. If the meniscus is too high, the block face will become wet, and the sections will drag down the back of the knife or lift off the water surface during the upstroke of the block.

13. With a coarse or medium microtome feed, bring the knife and block together carefully, using the microtome illumination, the microscope, and the reflection from the surface of the trough against the block face. Readjust the trough meniscus and the microscope until a silver reflection can be seen against the block face. The upper portion of the block face will appear silver; the lower portion, black [Fig. 3.12 (left)]. Bring knife and block together until the black portion is barely visible [Fig. 3.12 (right)]. You will be 3–10 μm from the block face at this point.

14. Readjust trough meniscus and illumination (see step 12) to the proper position for cutting, and move slowly toward the block (1–2 μm a stroke) until a section is cut (see Note a in following section.)

15. Advance 0.5–1 μm at a time manually (or on "automatic") at a cutting speed that is compatible with producing flat, chatter-free sections (1–2 mm per second).

16. When four or five sections have been cut, tease them to the center of the trough

with an ethanol-cleaned dissecting needle (being careful not to touch the knife edge) or with a clean coarse hair that has been taped to the end of an applicator stick (Fig. 3.13).

17. Pick up the sections, one at a time, with a clean platinum loop (see Note b in following section) that has been taped to the end of an applicator stick (Fig. 3.13) and transfer them to a clean slide. Alternatively, the sections may be picked up, one at a time, on a flat toothpick that has been sharpened to a chisel end (or with a clean dissecting needle) and placed in drops of water on a clean slide.

18. Place the slide on a hot plate at 80°C to flatten the sections and evaporate the water. The water should evaporate slowly so that no wrinkles or bubbles form; it is often helpful to lift the slide off the hot plate for a few seconds several times while the water is evaporating, especially during the final stage. Leave the slide on the hot plate for about 20 seconds after all the water has evaporated to ensure proper adhesion of section to slide. Prolonged heating can impede toluidine blue staining.

19. Put a few drops of 1% alkaline toluidine blue over the sections and return the slide to the hot plate for 10–60 seconds (see Note d in following section). (Staining time will depend upon the type of resin used: 10–20 seconds for Epon-Araldite; 60 seconds for Spurr's resin).

20. Wash the slide in fresh, hot tap water (cold water could crack the hot slide).

21. Drain off excess water by holding the slide vertically against a blotting tissue (Fig. 3.14A).

22. Remove the block from the chuck and label it with non-water-soluble ink. The block surface may first be roughened with a scalpel blade, but bear in mind that this causes fine polymerized resin dust to be released. Avoid inhaling this dust.

23. Put the labeled block into a labeled storage box (Fig. 2.2D).

24. Label the slide. When it has dried completely, coverslip it with resin that has been drawn into a 1-mL syringe fitted with a 22-gauge needle (Fig. 3.14B). These resin-filled syringes may be stored frozen but should be brought to room temperature before use to eliminate the possibility of moisture contamination. Discard (in a safe disposal container) when the resin becomes too viscous to spread under a cover slip. While this method has been found to give superior resolution (since the coverslip medium has the same refractive index as the embedding medium) commercial mounting media, such as Eukitt, dry quickly and might be more convenient for routine work. Commercial media do, however, usually cause section fading and wrinkling with time.

25. Keep the slides flat and at room temperature for a few days until the resin has partially polymerized, then seal the edges of the cover slip with nail polish (Fig. 3.14C). Heating the slide to hasten polymerization will cause the toluidine blue to fade.

Notes on Method

a. Some technologists prefer to approximate knife and block using a dry knife. In this case, a thin silver line will appear across the block face, and the upper portion of the block will appear black. Move the knife toward the block face until this silver line is reached.

(0.5 mm per second) usually produces the best results. Each section can be lifted from the knife edge with an ethanol-cleaned dissecting needle (as it is cut) and placed in a drop of water on a clean slide. If two or three sections are cut before picking them up, they may be guided away from the knife edge with a dissecting needle. (Forming a slightly convex meniscus on the trough will prevent the sections from sticking to the tape while attempting to pick them up.) The slide should be placed on a hot plate for 20–30 seconds to flatten the sections. The slide is then removed and the drop of water is drawn off from each section with a dissecting needle or a strip of blotting paper. This procedure eliminates the wrinkling that tends to occur when large sections are dried using only the hot plate. Reheating of the slide causes the sections to adhere.

g. Linear tissue (e.g., muscle, retina) will cut more easily if the fibers or layers run perpendicular to the knife edge.

h. Semithin sections should appear shiny on the surface of the knife trough. If they appear dull, a new, sharp area of the knife should be selected.

i. Figure 3.15 shows a section of eye tissue that has been cut from a block containing atmospheric moisture. The tissue was dehydrated and embedded and then left on a lab bench overnight instead of being placed in an oven. It can be seen that staining is not crisp, and the background resin, which should be clear or slightly pink, has also stained.

j. Epon or Epon-Araldite produce satisfactory sections at 0.5 μm, but Spurr's resin is best cut at 1 μm to avoid wrinkling and staining problems. Dense or fibrous tissue such as nerve or muscle often cut better at 0.75–1 μm.

k. Sections flatten better and have less tendency to form bubbles after coverslipping, if gelatin coated slides are used (method follows).

l. Reichert (now Leica) make a microtome accessory called a "mesa." It is comprised of a mirror (which can be adjusted to reflect the block face) fitted to an adjustable magnetic stand (Fig. 3.11B). The mesa was designed to facilitate trimming a pyramid using a glass knife but also saves a great deal of time when searching for discrete structures such as renal glomeruli. A block is trimmed using a dry knife and sections thick enough (1–2 μm) to produce surface topography. This is reflected in the mirror on the down stroke of the block, and an assessment can be made without the more time-consuming procedure described in the above method section.

Silver Staining of Semithin Sections (Rosenquist, Slavin, & Bernick, 1971).

Solutions

SATURATED ALCOHOLIC SODIUM HYDROXIDE (SODIUM ETHOXIDE)

1. Pour 100 mL ethanol into a plastic bottle (sodium hydroxide etches glass).
2. Add sodium hydroxide pellets to a depth of 0.5 in. (12.5 mm).
3. Secure bottle lid and shake well.
4. Age for 2 days before use (solution will turn brown). Solution keeps well.

SILVER NITRATE – 2%

1. Combine 2 g silver nitrate and 100 mL distilled water.

2. May be made and stored (in the dark) in a 6-oz (150-mL) prescription bottle or an amber glass bottle.

GELATIN – 3%

1. Add 3 g gelatin to 100 mL distilled water.
2. Dissolve using heat.
3. Make solution just before use and use at room temperature.

HYDROQUINONE – 1%

1. Combine 1 g hydroquinone and 100 mL distilled water.
2. Store in the dark.
3. Make fresh each day it is to be used.

DEVELOPING SOLUTION

1. Combine 40 mL of 3% gelatin and 10 mL of 2% silver nitrate.
2. Mix well and add 4 mL of 1% hydroquinone while stirring.
3. Prepare immediately prior to use.
4. Discard after use.

GOLD CHLORIDE – 1%

1. Combine 1 g gold chloride and 100 mL distilled water.
2. Store in the dark. (Solution keeps well.)
3. Dilute to 0.3% (30 mL of 1% gold chloride diluted to 100 mL) if toning occurs too rapidly to control easily.

OXALIC ACID – 2%

1. Combine 2 g oxalic acid and 100 mL distilled water.
2. Dissolved and store at room temperature.

SODIUM THIOSULFATE – 5%

1. Combine 5 g sodium thiosulfate and 100 mL distilled water.
2. Dissolve and store at room temperature.

Method

1. Cut 1-μm sections as described earlier in Semithin-Section Cutting Method, steps 1 to 21.
2. Pick up sections and transfer them to a clean but uncoated slide.
3. Flatten the sections and evaporate the water on an 80°C hot plate; leave the slide on the hot plate for approximately 10 minutes after the water has evaporated.
4. Deplasticize the sections by treating them for 25 minutes in sodium ethoxide. Use plastic Coplin jars for this and subsequent steps.
5. Rinse several times in absolute ethanol.
6. Wash thoroughly in tap water, followed by a distilled water rinse.
7. Stain 2 hours in 2% silver nitrate at 50°C.

8. Rinse 1 to 2 seconds in distilled water.
9. Develop in silver–gelatin–hydroquinone mixture until the sections turn dark brown (or control the development microscopically). Discard the solution after use.
10. Wash well in tap water to remove the hydroquinone.
11. Tone, one slide at a time, in a jar of 0.3% gold chloride for 3–5 seconds (sections turn grey).
12. Rinse in tap water: Check microscopically (sections should have black basement membranes against a nearly colorless background) and tone again if necessary.
13. Wash in tap water for 5 minutes.
14. Dip in 2% oxalic acid for about 4 seconds.
15. Rinse in tap water.
16. Treat with 5% sodium thiosulfate (hypo) for 5 minutes.
17. Wash in tap water for 5 minutes.
18. Air dry and apply cover slip with resin.

Notes and Method

- This is a particularly good stain for kidney tissue since it delineates basement membranes well (Fig. 3.1B). Mitochondria and striated muscle are also well stained by this method.
- Silver–gelatin–hydroquinone developer must be at room temperature, 20°C. (Above this temperature, the solution becomes black very quickly and leaves a background deposit on the slide.)

THIN SECTIONS

Introduction

Sections for the electron microscope are cut from a preselected area of the block, usually based on information provided by viewing a semithin section with a light microscope. Section thickness is determined by the interference colors seen through the microtome binocular microscope as the sections float on the surface of the knife trough. Grey sections are thinner than 60 nm and are useful for high-resolution work. Silver sections are 60 to 90 nm, and this is the thickness range generally used. Gold sections are 90 to 150 nm and are useful for low-magnification work. It is useful to collect some sections of each thickness on the same grid.

Diamond knives produce thin sections of the best quality but, with care and experience, quite acceptable thin sections can be cut using a glass knife. The knife, however, should be made the day it is to be used.

The most serious problems facing an ultramicrotomist are static electricity, vibrations, air currents, dust, grease, bacteria, and other contaminants. Running a humidifier in the cutting room (being careful to deflect the draft away from the microtome) and using a static eliminator can greatly reduce the problem of static electricity. The author also finds a static eliminator (Fig. 3.18) very helpful during the cutting of semithin sections. It might be necessary to seal off air vents near the microtome or to encase the whole microtome in a Perspex (plexiglass) hood. A three-sided cardboard draught protector (Fig. 3.21) placed around the knife area works well on some microtomes. Clean hands, clean tools, and a dust-free work area prevent specimen contamination and cutting difficulties due to knife-trough contamination. Vibrations come from many sources and must be eliminated if acceptable sections are to be cut for the electron microscope.

Since electrons will not penetrate glass slides, thin sections are placed on mesh grids. Several mesh sizes are available but some (150 mesh or larger) must be coated with a support film. A thin copper foil grid with 5-μm bars (instead of the standard 20 μm) is available, combining small mesh size (and therefore maximum specimen stability) with a large viewing area.

Before viewing in the electron microscope, thin sections of biological material must be stained with heavy metals. Although some contrast results from postfixation with osmium tetroxide and en bloc staining during dehydration, further staining is done after thin sectioning to obtain optimal contrast (Fig. 3.31). The standard method is application of uranyl acetate followed by lead citrate. In the method outlined, it was found that uranyl acetate made in 30% ethanol was the most satisfactory; 50% ethanol softened the resin enough to cause section wrinkling, and an aqueous solution did not stain intensely enough.

If the ultramicrotome is not equipped with a water-leveling device, a very acceptable one can be made from a 1-mL tuberculin syringe and a scalp vein infusion set (Fig. 3.16). The syringe should be taped to a convenient place on the ultramicrotome. The needle should be gently bent into a "U" shape with pliers and hung over the knife trough, well away from the knife edge.

Chemicals and Equipment

Thin-Section Cutting

- Water, sterile distilled (or preferably sterile deionized double-distilled water).
- Butterfly infusion set – 25-gauge needle (bent) (Fig. 3.16).
- Hypodermic syringes – 1, 20 mL.
- Hypodermic needles – 22 gauge, 38 mm (1.5 in.).
- Filters – 0.22-μm pore size (e.g., Gelman 4418) (Fig. 3.19).
- Draft protector (Fig. 3.21).
- Jeweler's forceps, Dumont steel – No. 7, fitted with 7-mm ID (9/16 in.) rubber "O" rings (Fig. 3.24B).
- Fine hairs – Fullam No. 54220 or rat's whisker taped to end of applicator stick (Fig. 3.24A).
- Copper grids – a suitable mesh or Gilder high transmission grids, which have 5-μm instead of 20-μm bars (Fig. 3.24C).
- Chloroform.
- Polypropylene tubes – 2 × 75 mm (Fig. 3.22).
- Cotton-tipped applicator sticks (Fig. 3.22).
- Ultrasonication detergent (e.g., Liquinox).
- Sodium hydroxide pellets.
- Diamond knife – 2.5–3.5 mm.

● Static eliminator, e.g., Model P2402 (NRD). (NOTE: Because this device contains polonium, some countries require a radioactive device license) (Fig. 3.18).
● Silicone pads for grid storage (Fig. 3.24C).

Thin-Section Staining

● Beakers — Tri-pour, 50 mL.
● Bibulous paper — 100 × 150 mm (4 in. × 6 in.).
● Volumetric flask — 50 mL.
● Amber glass bottles — 100 mL.
● Petri dishes — 60, 90 mm.
● Filters — 0.45 μm, CR (e.g., Gelman 4219).
● Sodium hydroxide pellets.
● Sodium hydroxide — 1 N, carbonate-free.
● Lead acetate [$Pb(C_2H_3O_2)_2 \cdot 3H_2O$].
● Sodium citrate ($C_6H_5Na_3O_7$).
● Uranyl acetate [$UO_2(C_2H_3O_2)_2 \cdot 2H_2O$].
● Ethanol, absolute.
● Hypodermic syringe — 20 mL.
● Jewelers forceps No. 5, Dumont steel fitted with "O" rings (Fig. 3.24B).

Grid-Coating

● Glass staining jar — 75 × 100 mm (3 in. × 4 in.).
● Formvar (0.25% or 0.5%, in ethylene dichloride) or Butvar.
● Chloroform.
● Note paper – coarse yellow scratch pad.

Pyramid-Trimming Method

1. Using a light microscope, select an area of interest on a semithin toluidine blue section.
2. Position the corresponding block in a specimen holder, then lock into a trimming block on the stage of the ultramicrotome. The area of interest should be visible on the face of the block when viewed through the ultramicrotome microscope at 15–20 ×.
3. Check the slide again and look for "landmarks" that can be recognized on the block face. Sometimes it helps to waft chloroform vapor over the block face.
4. With clean hands and a degreased, halved, double-edged razor blade, trim into the desired area by cutting toward you at a 50° slope and parallel to the block edge, taking only a thin sliver of plastic with each slice [Fig. 3.17 (top)].
5. Turn the block 180° so that the opposite edge is nearest, and reposition or refocus if necessary. Wipe the blade with ethanol and repeat Step 4, making certain that this edge is parallel to the first one trimmed.
6. Turn the block back to its original position. Wipe the blade again and trim the sides of the block, maintaining the 50° slope and trimming so that the bottom edge of the block (nearest edge) is somewhat longer than the top edge (farthest edge) [Fig. 3.17 (bottom left)]. While this shape is most desirable, a rectangular face will cut well

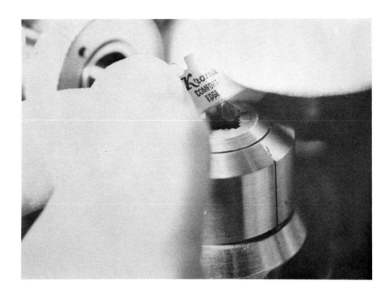

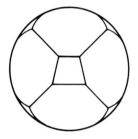

Figure 3.17. *Pyramid prior to thin sectioning.*

Top. Trimming a pyramid. Thin slivers of plastic should be removed with each slice. **Bottom left.** Ideal block face. **Bottom right.** Rectangular block face.

Figure 3.18. Static eliminator for use during cutting.

[Fig. 3.17 (bottom right)]. The block face should be 0.2 to 0.4 mm on a side and should not exceed 0.5 mm. Section quality deteriorates as block size increases.

7. Remove the block from the chuck, rinse it with 70% ethanol followed by sterile distilled water, and dry it thoroughly with compressed air before replacing it in the chuck.

NOTE ON METHOD. Every trace of packing grease must be removed from a razor blade before using it to trim a pyramid. The degreasing may be done with ethanol. The edge of the razor blade should be wiped with ethanol prior to trimming each of the four sides of the pyramid. Any contamination along the edge of the block or on the block face will cause cutting problems far more serious than those experienced by slightly nonparallel edges.

Thin-Section Cutting Method

It is recommended that glass knives be used until the microtomist feels very familiar with the ultramicrotome.

1. Clean the ultramicrotome with a damp sponge or cloth. This not only removes all traces of dust, lint, stray sections, or bits of plastic, but the moisture helps to reduce static electricity. Also clean the area around the microtome.
2. Wash hands thoroughly before starting cutting procedure as well as after touching one's face or hair, blowing one's nose, or doing anything that might contaminate one's hands in any way with grease or bacteria. Move well away from the ultramicrotome before coughing or sneezing.
3. Clean the forceps and teasing hairs with absolute ethanol.
4. Flush the water-leveling device with fresh sterile or filtered, distilled water. Cleanliness is of the utmost importance during each step in the process of obtaining and staining thin sections.
5. Position the static eliminator 1–2 in. (25–50 mm) from the block and knife (Fig. 3.18).
6. Position the block in the ultramicrotome and clamp tightly.
7. Position the knife in the knife holder, being certain that knife and block are retracted to the point where they cannot possibly strike each other. Clamp tightly.
8. Slightly overfill the knife trough with fresh sterile distilled water either from a beaker with a Pasteur pipette or, preferably, from a 20-mL cleaned syringe that has been fitted with a cleaned 0.22-μm filter (Fig. 3.19).
9. Clean the tip of the water-leveling device with absolute ethanol. Clear any bubbles from the tube, and position the device over the knife trough.
10. With the binocular magnification set at 10–15 ×, use the microtome illumination and water-leveling device to obtain a flat surface that will give a silver reflection; then adjust the block height according to the manufacturer's instructions.
11. Raise the magnification to 20–30 ×. Use a combination of chuck and knife alignment knobs to ensure that the bottom and top edges of the block are parallel to the knife (Fig. 3.20A), that the block face and knife edge are parallel to each other (Fig. 3.20B), and that the block face is in a truly vertical position (Fig. 3.20C). This can be done by slowly and carefully moving the block downward past the knife. The reflection should be the same width from bottom to top of the block (i.e., the distance between the knife and block are the same bottom to top).
12. Bring the knife and block together slowly and carefully, using first coarse and then fine advance, until the block face is nearly reached.
13. Switch the ultramicrotome to automatic and, using the fine feed, advance until sections begin to cut.
14. Pick up any unsuitable sections with a coarse hair (used when semithin sectioning)

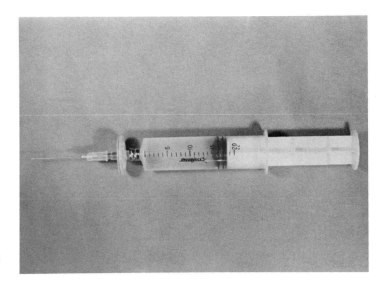

Figure 3.19. Syringe fitted with a 0.22-μm filter.

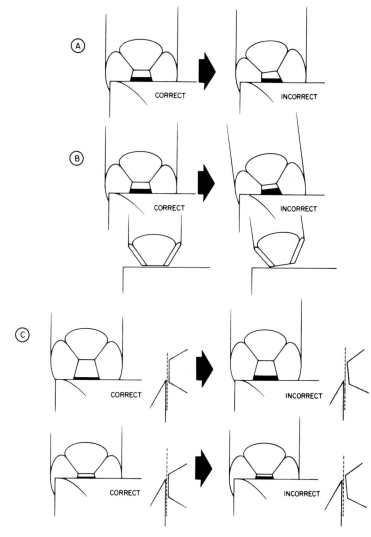

Figure 3.20. *Alignment of block and knife to cut thin sections.*

A. Bottom and top edges of block must be parallel to knife.

B. Block face and knife edge must be in same plane.

C. Block face must be in same plane as knife edge through entire cutting stroke.

Figure 3.21. Three-sided box of heavy cardboard used as a draft protector. (f) Front is $5\frac{1}{4}$ in. (132 mm) wide × 4 in. (103 mm) high; (s) is 4 in. (103 mm) × $6\frac{1}{4}$ in. (160 mm).

Figure 3.22. Chloroform used to flatten thin sections, with cotton-tipped applicator stored in polypropylene tube taped to bottle. (Xylene may also be used.)

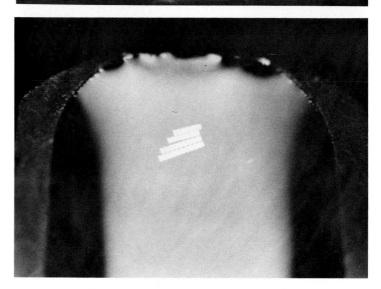

Figure 3.23. Thin sections arranged in a knife trough prior to being picked up on a grid.

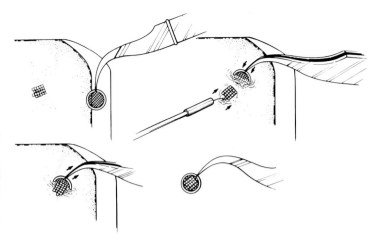

Figure 3.24. *Collection of thin sections.*
A. Collection of sections on a copper grid. **Top left.** Sections ready for collection. **Top right.** Grid put into the water at a 45° angle and sections teased toward it. **Bottom left.** Grid being lifted slowly out of the water, stretching the sections across it. **Bottom right.** Grid showing sections attached.

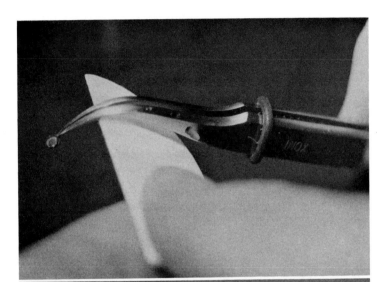

B. Blotting water from forceps.

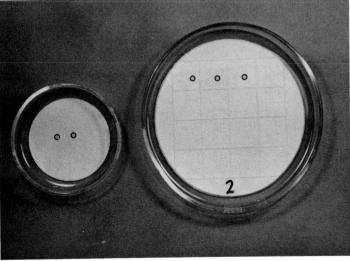

C. Grids of thin sections on filter paper or silicone prior to staining.

that has been cleaned with ethanol. Discard the first four or five sections (since these will be of inferior quality) unless there is a very restricted range of thickness available (such as an inclusion in a single cell).

15. Set the automatic thin-sectioning feed.

16. Put a draft protector in place (Fig. 3.21).

17. Cut 8 to 12 silver or pale gold sections (60–90 nm). For routine work, cut a range of sections from grey-silver (for high magnification) to pale gold. The latter will provide better contrast at low magnification.

18. Remove the draft protector.

19. Dip a cotton-tipped applicator in chloroform (Fig. 3.22) and waft the vapor over the sections, being careful to avoid any direct contact with the water surface. The sections should spread slightly but, if they spread much (e.g., change color from gold to silver), compression during cutting should be suspected. This can be caused by any or all of the following: (1) an incorrect knife angle (see Notes on Method), (2) poorly infiltrated and/or cured blocks, or (3) moisture in the embedding medium, picked up during preparation.

20. Using two fine hairs that have been cleaned in ethanol, tease the sections into two or three rows, each containing four or five sections (Fig. 3.23). If ribbons do not form, see "b" in Notes on Method.

21. Clamp a grid in No. 7 tweezers, clean it with a stream of ethanol from a wash bottle, and then rinse with a stream of distilled water (preferably from a syringe fitted with a 0.22-μm filter).

22. Hold the grid, dull side facing the operator, at approximately a 45° angle to the surface of the boat and immerse it in the water, almost to its upper edge. Tease the group of sections toward it with a fine hair (Fig. 3.24A).

23. Straighten the grid until nearly vertical and withdraw from the water.

24. Blot the water from between the arms of the forceps by running a folded filter paper from the base to the tip of the forceps and lightly touching the grid (Fig. 3.24B). Allow the grid to dry while clamped in the forceps. Sections should be on the dull or mat side of the grid where they are seen more easily than if collected on the shiny side.

25. When all traces of water are gone, place the grid, section side up, on a piece of labeled No. 1 filter paper (4.25 cm) in a 60-mm Petri dish or on a silicone pad in a 90-mm Petri dish (Fig. 3.24C). The grids are now ready to be stained.

26. If a diamond knife has been used, any sections left after a grid has been filled may be picked up on the end of a coarse hair, thereby forming a "mop." If this is run along the knife edge, it will pick up any section fragments that have adhered. This method is less damaging to the knife than using a wooden stick each time. (The water surface of the boat may be cleaned with cellulose tape. The surface should be level with the top of the boat and the tape long enough to hold with both hands. Draw it along the surface, being careful not to touch the knife edge. Never clear excess sections away with a dissecting needle as is sometimes done with a glass knife.)

27. Overfill the trough with distilled water to ensure sections do not adhere to a dry knife. Remove the knife. Wash it with a stream of 70% ethanol and/or a stream of distilled water from a wash bottle. Dry with compressed air in order to prevent

Figure 3.25. Electric knife cleaner.

growth of bacteria. If a glass knife has been used, discard it or keep it for trimming prior to cutting semithin sections.

Notes on Method

a. Gilder high transmission grids, while having the advantage of 5-μm rather than the usual 20-μm grid bars are very thin and therefore very fragile. They must be handled carefully to avoid bending, but since they permit 80% transmission, they are probably worth the extra trouble.

b. If cutting fails to produce ribbons (due to static electricity, contamination along the block edge or on the knife edge, the block not having parallel edges, or for any other reason), the following procedure is used.

   b1. After cleaning and blotting the grid, place it on a piece of No. 42 (more lint free than No. 1) filter paper. Put one drop of filtered water beside the grid, allowing the water to seep under the grid.

   b2. Flame the 3-mm platinum loop used for semithin sectioning. Wash it in ethanol, then in filtered distilled water, and blot dry.

   b3. Using fine hairs, tease the free-floating sections into as tight a cluster as possible, well away from the edge of the knife.

   b4. Lower the platinum loop over the sections and just make contact with the water. Do not go near the knife edge.

   b5. Lift the loop (which now contains the sections) and lower it carefully onto the grid. The water will be pulled away by the damp filter paper and the sections will flatten onto the grid. Then the loop may be lifted away.

c. Some diamond knives may be cleaned in an ultrasonic cleaner. If a 50-mL tri-pour beaker is used (Wallstrom & Iseri, 1972), the knife edge cannot come into contact with the sides of the beaker. Normally, 7% Liquinox (or other detergent recommended for ultra-sonication) in distilled water is recommended as the cleaning solution. Alternatively, the knife may be soaked in 7% Liquinox in a tri-pour beaker

1 hour to overnight, then flushed well with sterile distilled water or cleaned in a device which has been designed for this purpose (Fig. 3.25).

d. Should a diamond knife become very contaminated with adherent sections, clean it with a stick of balsa wood or styrofoam that has been sharpened to a fine chisel edge and soaked in 6 N (24%) sodium hydroxide. This must be done very carefully, permitting the sodium hydroxide to touch only the diamond, remain briefly, and then be washed away thoroughly with distilled water. The stick should never move at right angles to the knife edge (always parallel to it), and the cleaning should be done with the aid of a dissecting microscope.

e. If a diamond knife fails to become wet, this difficulty can usually be overcome by cleaning the knife or by filling the boat with 0.5% Aerosol OT (Garner & Steever, 1970) or 7% Liquinox, allowing it to stand for 15 minutes to an hour, then rinsing well with distilled water.

f. Should any grease, dust, or other contamination be in evidence on the surface of the boat, discard the knife if it is glass. If it is a diamond knife, remove it, clean it, and start again.

g. If sections fail to form ribbons or if they scatter rapidly around the surface of the boat when chloroform vapor is wafted over them, this is almost certainly due to a contaminated knife or to unclean edges on the block pyramid.

h. Knife-angle adjustment will vary from one microtome to another, but the angle is usually from 2° to 6° and must be determined by trial and error. Start with an angle of 4°–5° and check the sections for compression or chatter using the electron microscope. Compression indicates that the angle should be increased, and chatter indicates that the angle should be reduced. If gold sections are cut, they can be checked without staining (since they have been stained en bloc). Mark the optimum angle for each diamond knife on its box.

i. A cutting speed of 1–2 mm per second allows sections to flatten well and is recommended when using a diamond knife.

j. Blocks must be held firmly in their chucks, with no more than a few millimeters projecting beyond the chuck. The sides of the pyramid should slope at approximately 50° to provide maximum stability. All ultramicrotome screws must be tightened securely.

k. The following is a color guide to approximate section thickness:

Grey	< 60	nm
Silver	60–90	nm
Gold	90–150	nm
Purple	150–190	nm
Blue	190–240	nm

l. Uncoated grids may be made more adhesive by applying 0.5% Formvar in ethylene dichloride. Put the grids, dull side up, on a filter paper and apply a drop of Formvar to each one. Allow to dry.

m. Placing grids, section side up, within 50 mm (2 in.) of an ultraviolet lamp for 5 minutes prior to staining, helps sections stick to the grid during staining.

n. Gelman filters must be flushed out with ethanol (if No. 4129 Acrodisc CR) or water (if No. 4217 or No. 4418) to remove surfactant that would contaminate the grids. Fill a 20-mL syringe, turn it upright, and clear any air. Fit the filter and flush.

Figure 3.26. *Petri dishes prepared for staining.*

Left. Dental wax sitting on moist bibulous paper for uranyl acetate. **Right.** Dental wax surrounded by sodium hydroxide pellets for lead staining.

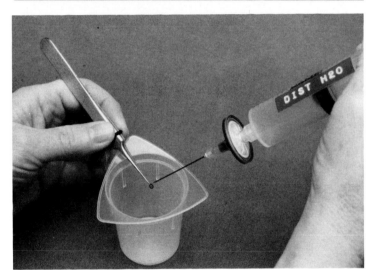

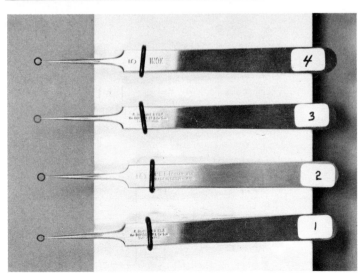

Figure 3.27. Grids being washed (top) and dried (bottom) after staining.

o. Hypodermic syringes can contain grease. Plungers should be wiped clean and barrels flushed with acetone. Allow to dry prior to use.

Thin-Section Staining

Solutions

URANYL ACETATE

1. Combine 0.6 g uranyl acetate and 20 mL 30% ethanol (prepared with sterile distilled water).
2. Mix well, filter, and store in an amber glass bottle (or clean 20-mL hypodermic syringe) in the dark. (Shelf life is 1–3 weeks.)

REYNOLDS' LEAD CITRATE (MODIFIED) (REYNOLDS, 1963)

1. Combine 1.33 g lead acetate and 1.76 g sodium citrate in a 50-mL volumetric flask.
2. Add 30 mL sterile distilled water.
3. Shake resultant suspension vigorously for 1 minute and allow to stand for 30 minutes with intermittent shaking to ensure complete conversion of lead acetate to lead citrate.
4. After 30 minutes, add 8 mL of $1 N$ sodium hydroxide (carbonate-free).
5. Dilute to 50 mL with sterile distilled water and mix by inversion. When lead citrate dissolves, the solution is ready for use.
6. The solution should be kept where it will not be disturbed (Fig. 3.28). It should be discarded when it becomes slightly milky or after 1 month.

Method

1. Put a folded piece of bibulous paper in the bottom of a 90-mm $(3\frac{1}{2}$ in.) Petri dish [Fig. 3.26 (left)].
2. Thoroughly dampen the bibulous paper with a stream of sterile distilled water. If possible, stain near a sink with the hot water running. Keeping atmospheric moisture high prevents surface drying of the stain.
3. Place a piece of dental wax on top of the bibulous paper (do not touch the upper surface of the dental wax).
4. On the dental wax, place pools of freshly filtered uranyl acetate to equal the number of grids to be stained. Each pool should contain 3–4 drops of stain. Cover immediately. The uranyl acetate may be filtered onto the wax using No. 1 filter paper and a clean funnel. Discard the first two or three drops into a waste container. Preferably, uranyl acetate may be filtered through a flushed 0.45-μm pore size Acrodisc CR filter attached to a hypodermic syringe fitted with a 21-gauge needle. (See Notes on Method "b" in following section.) The first two or three drops of stain should be expelled into a waste container, and the tip of the needle wiped with a tissue.
5. Lift the cover and float one grid, section side down, on each pool of uranyl acetate. Replace the cover of the Petri dish after positioning of each grid, since it is important to reduce surface drying of uranyl acetate as much as possible.

Figure 3.28. Lead citrate in protected area on the lab bench where it will not be disturbed or moved.

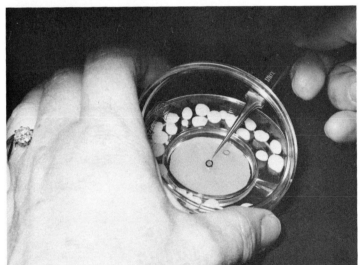

Figure 3.29. Shielding lead stain from CO_2 vapors.

6. Stain for 8–10 minutes.
7. After staining, lift the lid, clamp a grid in a pair of No. 5 Dumont forceps, and quickly (before any uranyl acetate can dry) wash well in a stream of filtered 30% ethanol (3–5 mL) [Fig. 3.27 (top)]. Let the stream of ethanol run down the arms of the forceps, flowing over the grid. It is important to remove all traces of uranyl acetate. Blot with a folded piece of bibulous paper, as shown in Fig. 3.24(B), running it from the base to the tip of the forceps, lightly touching the edge of the grid. Place

Figure 3.30. *Micrographs showing advantage of using sterile distilled water.*
A. Section of muscle showing contamination by breakdown products of bacteria.

B. Section from same muscle tissue (A) but using filtered distilled water to prepare stains and washing solutions.

the forceps holding the grid (section side up) in a clean area to dry [Fig. 3.27 (bottom)].

8. Repeat step 7 for each grid, closing the lid of the Petri dish after removing each grid. Leave the grids until they are completely dry before staining in lead citrate. Residual ethanol from the uranyl acetate stain may cause enough turbulence, when mixed with water from the lead citrate stain, to dislodge the sections.

9. Wash the dental wax with 30% ethanol and dry. Discard the wet bibulous paper from the Petri dish unless it is to be used within the day. Mold and bacteria can develop if wet paper is left in the covered dish.

10. Place a piece of dental wax in the bottom of a Petri dish and surround it with sodium hydroxide pellets [Fig. 3.26 (right)]. Replace the lids of both the Petri dish and sodium hydroxide pellet bottle. Sodium hydroxide is used to absorb atmospheric

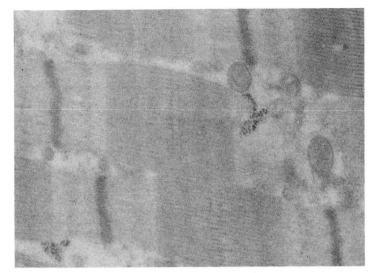

Figure 3.31. *Comparison of staining methods.*
A. En bloc: uranyl acetate; grid: unstained.

B. En bloc: uranyl acetate; grid: lead citrate.

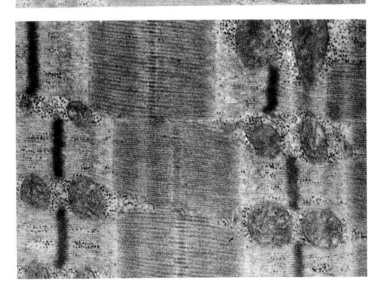

C. En bloc: uranyl acetate; grid: uranyl acetate, lead citrate.

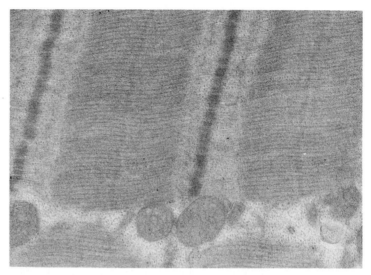

D. En bloc: lead acetate, uranyl nitrate; grid: unstained.

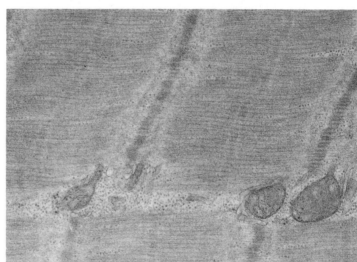

E. En bloc: lead acetate, uranyl nitrate; grid: lead citrate.

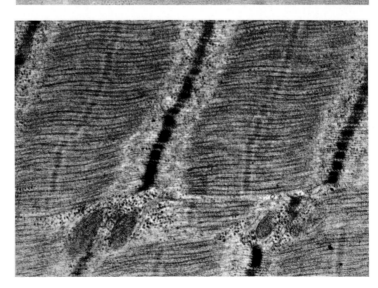

F. En bloc: lead acetate, uranyl nitrate; grid: uranyl acetate, lead citrate.

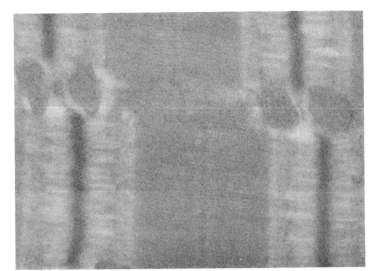

G. En bloc: uranyl nitrate; grid: unstained.

H. En bloc: uranyl nitrate; grid: lead citrate.

I. En bloc: uranyl nitrate; grid: uranyl actate.

CO_2, which would react with lead citrate to cause a lead carbonate precipitate on the grids ($2NaOH + CO_2 \rightarrow Na_2CO_3 + H_2O$).

11. Being careful not to disturb any precipitate, remove the stopper from the previously prepared volumetric flask of lead citrate solution (Fig. 3.28) and pipette a small amount (20 drops) of stain from just beneath the surface. Replace the stopper immediately.

12. Expel the first two or three drops of stain into a waste container, then place one drop for each grid to be stained on the dental wax. Replace the Petri dish lid. Do not breathe over the Petri dish while the stain is exposed. The CO_2 content of expired air is very high. (Keeping the humidity high will also help reduce atmospheric carbon dioxide.) Figure 3.29 shows how the grids can be handled while shielding the lead stain.

13. Float one grid, section side down, on each drop of lead citrate, replacing the lid after positioning each grid. Stain for 3 minutes.

14. After staining, quickly wash each grid thoroughly, as in step 7, with filtered distilled water. Blot and dry also as in step 7.

15. Remove, wash and dry the dental wax and return to the Petri dish. Replace the Petri dish lid. The sodium hydroxide pellets must be discarded when they appear wet.

16. When the grids are completely dry, place them (section side up) in a Petri dish on filter paper or on a silicone pad. They are ready to be viewed with the electron microscope.

Notes on Method

a. Uranyl acetate and lead citrate washings should be disposed of in sealed waste containers and not thrown down the sink.

b. Gelman Acrodisc CR filters must be flushed with absolute ethanol to remove the surfactant, then sufficient uranyl acetate (1–2 mL) forced through to replace the ethanol. The filter should be changed once a day to ensure clean grid preparations. A 5.2-mm BEEM capsule fits the end of a Gelman filter snugly. This will help prevent the stain from drying in the filter. Uranyl acetate must be kept in a dark place so store the syringe and filter in a covered box or in a refrigerator. Lead citrate may be stored in the volumetric flask used for its preparation, in a stable area where it will not be disturbed (Fig. 3.28), or it may be drawn into a syringe and dispensed through a (clean) filter in the same manner as uranyl acetate.

c. Sterile distilled water should be used whenever thin sections are cut or stained and whenever stains or rinsing solutions are prepared. Breakdown products of bacteria stain with heavy metals and can contaminate sections. Figure 3.30A shows such a contaminated section. Figure 3.30B shows a section from the same material, cut at the same time, but the stains and washing solutions were all prepared with sterile distilled water.

d. Nalgene wash bottles, tubing, and storage bottles should be cleaned frequently with sodium hypochlorite (household bleach), since bacteria collect on their walls. The sodium hypochlorite must be rinsed away thoroughly.

e. Figure 3.31A–I is a comparison of staining methods that illustrates the benefit of staining thin sections on the grid in addition to en bloc staining during processing. The tissue was embedded in Spurr's resin.

Grid-Coating Method

Although support films reduce contrast in the electron microscope and present a source of contamination, it is sometimes essential that they be used. Grids larger that 150 mesh will usually require a support film.

1. Wash hands before beginning procedure.
2. Fill a clean staining jar to the top with sterile distilled water. Clean the surface by drawing cellulose tape across it. The tape should be long enough to grasp with both hands, keeping fingers outside the staining jar.
3. Wipe a 25 × 75 mm (1 in. × 3 in.) glass slide free of dust, but do not clean it well or the Formvar film will not float off easily.
4. Dip the slide until it is three-quarters immersed in 0.25% (or 0.5%) Formvar in ethylene dichloride and withdraw it again promptly. Commercially prepared Formvar solution is best. If it does not come in a bottle with a neck wide enough to accept a glass slide, a Coplin jar may be used. Expose the Formvar solution to the atmosphere as little as possible since ethylene dichloride evaporates very rapidly and the solution is hygroscopic.
5. Drain the slide in an upright position on absorbent paper and allow to dry 3 or 4 minutes.
6. Score around the bottom three-quarters of the slide with a clean razor blade.
7. Breathe on the slide (to create moisture) and immediately dip it into the staining jar at an angle of 25° to 30°. The thin layer of plastic should detach itself and float on the surface as the slide is lowered into the water. Do not allow your fingers to touch the water as this will cause contamination.
8. Look along the surface of this film (with the aid of a lamp if necessary) and, choosing only grey or silver areas that look free of flaws, place several grids, dull side down, on top of the film.
9. Cut a piece of coarse scratch-pad paper the same size as a slide. Make a fold at one end to form a handle.
10. Place the paper on top of the grids.
11. When the paper has been thoroughly moistened, lift and place it to dry grid side up, on paper toweling or other absorbent surface in a clean Petri dish.

Notes on Method

- Typing or note-pad paper is not porous enough to ensure firm adhesion of grids and Formvar, whereas filter paper is too porous and sinks unless it is supported by screening. Inexpensive yellow scratch-pad paper is recommended.
- Butvar (0.25% in chloroform) is an alternative to Formvar. This solution must be made fresh and discarded after use.

REFERENCES

Garner, G. E., & Steever, R. G. E. (1970). Treatment of diamond knives with Aerosol OT for uniform wetting of their edges in ultramicrotomy. *Stain Technol.* **45**: 186–187.

Reynolds, E. S. (1963). The use of lead citrate at high pH as an electron opaque stain in electron microscopy. *J. Cell Biol.* **17**: 208.

Rogers, A. S. (1973) *Techniques of Autoradiography*. Amsterdam: Elsevier Scientific, p. 108.

Rosenquist, T. H., Slavin, B. G., & Bernick, S. (1971). The Pearson silver gelatin method for light microscopy of 0.5–2 µm plastic sections. *Stain Technol.* **46**: 253–257.

Trump, B. F., Smuckler, E. A., & Benditt, E. P. (1961). A method for staining epoxy sections for light microscopy. *J. Ultrastruct. Res.* **5**: 343.

Wallstrom, A. C. & Iseri, O. A. (1972). Ultrasonic cleaning of diamond knives. *J. Ultrastruct. Res.* **41**: 561–562.

4

• *Immunoelectron Microscopy*

INTRODUCTION

In the early 1940s, immunocytochemistry was introduced for the immunodetection of antigens on tissue sections at the light microscope level. This innovative approach led to the development of various variants in cytochemical techniques with improved versatility and resolution and their extended application in all fields of biological research and diagnostic pathology. Adaptation of cytochemistry to the electron microscope level led to a second (with the immunoperoxidase techniques) and a third wave of interest (with the immunogold techniques), in the 1970s and 1980s respectively, with the application of these techniques to a large variety of research and diagnostic activities. The colloidal gold marker was first introduced in immunocytochemistry by Faulk and Taylor in 1971 for the detection of membrane antigens using colloidal gold-tagged immunoglobulins. Since then, colloidal gold has become the electron-dense marker of choice in cytochemistry because of the several and major advantages it displays when compared to other tracers such as peroxidase and ferritin. It can be prepared in very small sizes and, being electron dense, it is easily recognizable yielding labeling of very high resolution. It allows for accurate identification of the labeled structures without any masking effect. Being particulate, it can be quantitated, bringing an additional dimension to cytochemistry. It also offers the possibility of detection at the light microscope level. Preparation of colloidal gold-tagged molecules is a relatively easy task, and the procedure does not alter significantly the biological properties of the tagged molecule. Furthermore, it has little spontaneous affinity for the resins, giving by itself very low levels of background. As already mentioned, immunoglobulin–gold complexes were the first probes in immunogold cytochemistry to be applied to pre-embedding procedures. Later, in 1977 and 1978, the protein A–gold was introduced as an advantageous alternative to the immunoglobulin–gold for the pre- (Romano & Romano, 1977) and post-embedding (Roth, Bendayan, & Orci, 1978) detection of tissue antigens at the electron microscope level with very high resolutions. Protein A, a constituent of the *Staphylococcus aureus* cell wall, has the

Department of Anatomy, University of Montréal.

particular property to interact with high affinity with the Fc fragment of immuno-globulins G from various mammalian species (Langone, 1982). This interaction, together with the fact that the approach is carried out in two steps, provides versatility to the protein A–gold approach. Following the development of the protein A–gold, alternative reagents have been introduced recently. Protein G, a constituent protein from the cell wall of group G streptococcal strains, has been found to display properties similar to protein A with the advantage of having higher affinities for immunoglobulins G of mouse and goat (Åkerström & Björck, 1986; Bendayan & Garzon, 1988). Protein A/G, a recombinant protein displaying the combined affinity properties of protein A and protein G, was generated recently (Eliasson et al., 1988), and the protein A/G–gold complex was demonstrated to be highly versatile and a very convenient probe for immunochemical and immunocytochemical techniques (Ghitescu, Galis, & Bendayan, 1991). In the present review, we will try to describe the technical procedures leading to the production of the colloidal gold particles, the fabrication of the protein A–gold complex, as well as the different steps in the post-embedding indirect immunogold labeling protocol for electron microscopy. Extended reviews (Roth, 1982; De Mey, 1983; Bendayan, 1984, 1989; Kellenberger & Hayat, 1991) on the subject have been published although with less emphasis on technical descriptions. These reviews are recommended for readers desiring a more complete survey of the literature.

PREPARATION OF THE COLLOIDAL GOLD

Several techniques have been published that describe the preparation of colloidal gold suspensions of different sizes (Handley, 1989). Among them, the technique of Frens (1973), using tetrachloroauric acid and sodium citrate as the reducing agent, certainly appears as the most reliable, easy to carry out and yields monodispersed particles very homogeneous in size. Further, it also allows for varying the size of the particles to work with, giving some latitude and opportunity to choose the appropriate size for the study. The technique is simple and variation in particle size is obtained by changing ratios between the sodium citrate and the chloroauric acid.

Method

The preparation of a monodispersed 15-nm gold particles suspension is carried out as follows.

1. Prepare a 100 mL solution of 0.01% $HAuCl_4$ in a 300-mL Erlenmeyer by adding 0.5 mL of a 2% $HAuCl_4$ stock solution to 99.5 mL of double-distilled water.
2. Cover the flask with tin foil to avoid extensive evaporation.
3. Bring this solution to the boiling point on a hot plate or over an open flame.
4. As soon as the solution boils and while keeping it boiling, add rapidly 4 mL of a 1% trisodium citrate ($Na_3C_6H_5O_7 \cdot 2H_2O$) solution.
5. Cover the flask with tin foil and keep the solution boiling for about 10 minutes till it becomes bright red (the characteristic wine-red color of the colloidal gold). The clear solution initially evolves as a grey-purple color, which then turns dark purple and finally develops into the bright red color characteristic of the colloidal gold. Once the colloid has formed, prolonged heating will not produce any changes in

the size of the particles. However, addition of electrolytes will induce flocculation, the colloidal suspension turning into a black-purplish color. Flocculated colloidal gold precipitates rapidly and is unable to form any protein-gold complex.

6. Remove from heat and chill.

Notes on Method

The method described above will yield monodispersed colloidal gold particles of about 15 nm in size. To obtain gold particles of a different size, the amount of sodium citrate added to the solution of tetrachloroauric acid should be varied. Increasing the amount of sodium citrate will result in a reduction in the size of the particles down to about 10 nm. By decreasing the amount of sodium citrate, the size of the particles will increase going up to 30 to 40 nm. In such a case, the length of time that the solution has to boil before reaching the red color can extend to 20 minutes.

Keeping the size constant from one preparation to another and making the technique a reproducible one requires some precautions and rules. This reproducibility is important, particularly if quantitative evaluations of densities of labeling are going to be performed.

- All glassware, particularly that in which the gold suspension is made and stored, should be scrupulously cleaned. Siliconization is not absolutely required. For example, we soak our glassware in nitric acid before washing in a Miele dishwasher using the Neodisher A8 with the neutralizor Neodisher N. After washing, the glassware is further thoroughly rinsed with double distilled water. Any impurity present on the glass surface will modify and alter the final result. It will either change the size of the particles or will simply induce flocculation of the gold suspension. We also recommend reserving a set of glassware to be used *exclusively* for the preparation of the colloidal gold.
- Since the size of the gold particles depends on the ratio between concentrations of tetrachloroauric acid and sodium citrate, we suggest using stock solutions of these reagents. The stock solutions can be prepared in large volumes that, if properly stored, will last for a couple of years. The tetrachloroauric acid can be prepared as a 2% stock solution by dissolving 1 g of $HAuCl_4$ in 50 mL of double distilled water. This solution can be kept at 4°C protected from light (wrapped with tin foil) for several months. It is quite important to prepare such a stock solution of the tetrachloroauric acid since this reagent is extremely hygroscopic, and it is difficult to weigh properly small amounts of this compound for each preparation of colloidal gold. The 1% solution of sodium citrate can be prepared in large amounts, (500 to 1,000 mL), which should be frozen in small aliquots (10 to 50 mL) and used as needed. By always measuring the *same* volumes of the *same* stock solutions reproducible results are assured in terms of size of the gold particles.
- Preparation of a 100-mL suspension with the indicated reagents will result in particles of the appropriate size. However, increasing by several fold the amount of each reagent will alter the final size of the colloidal gold particles. Reproducibility is not maintained when large volumes are prepared.
- It is important to prepare the 100 mL of the colloidal gold in a 300-mL Erlenmeyer flask in order to have a very large base exposed to the heat.
- Freshly prepared colloidal gold suspension should be used for the preparation of

protein–gold complexes. Any remaining suspension can be stored at 4°C protected from light for a couple of days. We do not recommend storing this for very long since aggregation of the colloid will occur.

PREPARATION OF THE PROTEIN A–GOLD COMPLEX

The conjugation of a protein with colloidal gold particles occurs upon contact when proper conditions of pH and concentration are provided. If conditions are inadequate, instead of a stable complex, flocculation takes place with precipitation of the gold. Thus, prior to the mixing of protein A with any colloidal gold suspension, the pH should be adjusted to the correct value. For protein A the optimal pH for adsorption onto gold particles has been found to be between 5.8 and 6.0 (Ghitescu & Bendayan, 1990). Adjusting the pH of the gold suspension is carried out with 0.1 N HCl or 0.2 N K_2CO_3. Since we are dealing with a colloidal suspension, measurement of the pH could be a problem. Indeed, a standard pH meter electrode should not be immersed into a colloidal suspension, since with time it will get coated with gold. Three alternative methods are available to measure pH values of colloidal gold:

1. Use of a gel-filled combination electrode specially designed for this purpose.
2. Estimation of the pH with a pH paper of good sensitivity (which in certain cases such as for protein A is quite adequate).
3. Use of a technique in which aliquots of colloidal gold are stabilized prior to reading the pH with a standard electrode. In this last technique, the aliquot of colloidal gold (5 mL) is mixed with 0.05 mL of 1% aqueous solution of polyethylene glycol (mw $\sim 20,000$) before reading the pH. This aliquot is then *discarded* and drops of either 0.1 N HCl or 0.2 N K_2CO_3 are added to the main colloidal suspension in order to reach the desired pH value. The pH reading procedure is repeated as many times as needed till the required pH value for the main suspension is reached. Care should be taken not to add the polyethylene glycol solution to the main suspension since it will stabilize the colloidal gold, preventing subsequent adsorption of protein A.

For the preparation of the protein A-gold complex (Ghitescu & Bendayan, 1990), 150 μg of protein A dissolved in 0.2 mL of distilled water are introduced in an ultracentrifuge tube, and 10 mL of the colloidal gold suspension (15-nm size) is poured rapidly while mixing both reagents. The order of addition of the two reagents is important; the gold sol should be added to the protein solution to prevent the formation of aggregates. Development of the protein A–gold complex occurs upon mixing, and a simple ultracentrifugation is sufficient to recover the protein A–gold complex, separating it from non-bound protein A. The ultracentrifuge tubes used are made of polystyrene; they may be either disposable or nondisposable. Centrifugation should be carried out without delay at 4°C using a fixed-angle Beckman Ti50 rotor and a Beckman ultracentrifuge at 25,000 rpm for 30 minutes. For smaller gold particles, the amount of protein A differs (60 μg), and the centrifugation should be carried out for longer periods (60 min) at higher speed (40,000 rpm). For larger particles (> 30 nm), the amounts should be increased and the centrifugation carried out at lower speed (15,000 rpm) for 30 minutes. After centrifugation, the tubes should be carefully removed

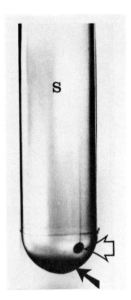

Figure 4.1. Three phases clearly visible in tube after centrifugation. (1) Clear supernatant (s) containing free protein A. (2) Dense precipitate (*large clear arrow*) at bottom of tube, corresponding to gold nonstabilized by protein A. (3) Sediment (*black arrow*) at bottom of tube representing protein A–gold complex.

from the centrifuge and the protein A–gold complex recovered without delay. Three phases are found in the centrifuge tube (Fig. 4.1): (1) a supernatant, (2) a dense precipitate, and (3) a sediment. The supernatant is clear and contains the free protein A. The precipitate is dark and usually adheres to the bottom on the side of the tube; it represents gold that was not stabilized by the protein. The red sediment corresponds to the protein A–gold complex. The supernatant is carefully aspirated as completely as possible and discarded. Since it contains free protein A it should be totally removed. Any contamination of the protein A–gold with free protein A will lower and hamper the efficiency of the probe, since free protein A will compete with protein A–gold for the Fc binding sites during labeling. When aspirating this supernatant, it is advisable to proceed until the red sediment is reached, and a small amount of this should be drawn up with the last trace of supernatant. The unstabilized gold remains at the bottom of the tube and will be removed during the washing of the tube. For this reason, it is important to use a fixed-angle rotor rather than a swinging-bucket system in which the unstabilized gold precipitate and the protein A–gold sediment will settle together, making it more difficult to separate them at the end of the centrifugation step. The protein A–gold sediment (about 0.5 mL) is recovered and resuspended in 1.5 mL of 0.01 mol/L phosphate buffered saline (PBS: NaH_2PO_4/Na_2HPO_4, NaCl 0.14 mol/L) pH 7.3 containing 0.02% polyethylene glycol (mw $\sim 20,000$) and stored at 4°C. At the labeling step, this solution is further diluted 5- to 10-fold with the same buffer to obtain a solution with an adsorbance reading of 0.5 to 1 at 525 nm.

Again, to assure reproducibility all glassware in which the complex is formed and stored should be scrupulously cleaned. The protein A–gold should not be frozen; when stored at 4°C under sterile conditions, its activity remains practically unchanged for several months. If freezing is absolutely required, the complex should be diluted in 50% glycerol (Slot & Geuze, 1981). For lyophilization, the complex should first be dialyzed against 5 mmol/L NH_4HCO_3 prior to lyophilization (Baschong & Roth, 1985).

LABELING PROTOCOL

The procedure to be described is the one required for transmission electron microscopy. Without going into the details of tissue preparation for electron microscopy, which are found in other chapters of this book, some elements are important enough to be mentioned because of their direct effect on the immunocytochemical results.

The goal of immunocytochemistry consists of identifying specific compounds and assigning them to particular tissue and/or cellular structures. For such purposes, two criteria are required: (1) retention of the biological properties of the chemical compound to be able to reveal it, and (2) preservation of the structural elements in order to identify them. It happens that both requirements are usually not compatible. Indeed, procedures and reagents that preserve the structural elements modify significantly the chemical configuration of tissue and cellular components. For optimal results, one has to compromise between the two criteria and work out the best combination that will allow preservation of biological activity with sufficient retention of structural elements. Preservation of tissue and cellular structures is achieved through two procedures – fixation and embedding. Each procedure influences the chemical nature of the components in a specific manner.

1. Fixation can be carried out with several chemicals, the most usual being glutaraldehyde, paraformaldehyde, or a combination of both. In order to retain chemical properties, the concentrations of the fixatives should be kept as low as possible (0.5–1% glutaraldehyde, 2–4% paraformaldehyde, or the combination 4% paraformaldehyde + 0.1% glutaraldehyde, as well as 4% paraformaldehyde-lysine-periodate). Fixation is carried out at room temperature for 2 hours or at 4°C for longer periods of time. When fixation is performed with paraformaldehyde alone, it should be carried out for long periods (over 24 hours), and washing with buffer should be reduced to a very short time since this fixation is reversible. For several antigens postfixation with osmium tetroxide (1% for 2 hours at 4°C) can be performed without hindering retention of antigenicity and further immunolabeling. Cryofixation versus chemical fixation can be of importance for the retention of certain antigens and also to avoid displacement of soluble ones. The fixatives can introduce and generate free aldehyde groups in the tissues that should be quenched in order to reduce background staining at time of labeling. This is performed by incubating the tissue blocks with 0.15 mol/L glycine or lysine in PBS or with 0.5 mol/L NH_4Cl solutions prior to proceeding with dehydration and embedding.

2. Embedding can be performed in various resins following different dehydration and infiltration methods that influence the preservation of antigenicity. Among numerous variants, five embedding methods should be mentioned: (1) epoxy resins (Epon, Araldite, and Spurr's resin) (Bendayan, 1984; Bendayan et al., 1987), (2) methacrylate resins (glycol methacrylates with or without dehydration in organic solvents) (Bendayan, 1984; Bendayan, Nanci, & Kan, 1987), (3) Lowicryl resins (K4M, K11M, HM20, HM23) (Bendayan, 1984; Bendayan et al., 1987; Hobot, 1989; Villiger, 1991), (4) LR resins (LR White or LR Gold) (Bendayan et al., 1987; Newman, 1989), and (5) no embedding in resin but rather processing by cryoultramicrotomy (Tokoyasu, 1983; Bendayan et al., 1987; Van Bergen en Henegouwen, 1989). The latter method is particularly useful for sensitive antigens since it avoids dehydration in organic solvents and infiltration in resins.

3. Sectioning of the tissue blocks should be carried out with care to avoiding artifacts

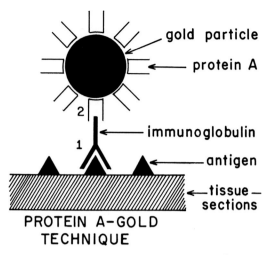

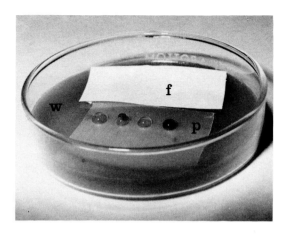

Figure 4.2 (left). Diagram illustrating the principles of postembedding protein A–gold technique, which is carried out in two steps: (1) Immunoglobulin interacts with its specific antigen molecule present at surface of section. (2) Protein A, present at surface of gold particle, binds to Fc fragment of immunoglobulin. By superposition, gold particle allows for indirect localization of antigenic site. (Proportions in diagram have not been respected) (from Bendayan, 1984).

Figure 4.3 (right). Incubation setup. Using petri dish coated with layer of dental wax (w), drops of different reagents are placed on a piece of Parafilm (p). Filter paper (f) soaked with water is placed nearby to assure a humidified atmosphere. During incubation, petri dish is closed. As shown, grids are incubated (sections face down) floating on the drops.

that may subsequently interfere with the labeling process. The sections should be mounted on nickel grids, which should be coated with a supporting film. The nature of the film has no influence on the labeling provided that, as mentioned later, the various reagents do not have access to the opposite side of the grid and thus to the film itself. The film provides good support for the sections, which is certainly needed at the time of washing during the labeling protocol. Without a supporting film the sections may be lost during incubation and washing procedures. The grids can be stored for several months. Protected from dust and humidity, the "old" sections will yield labelings comparable to those obtained with freshly prepared ones. Similarly, we found that antigenicity is not reduced by the storage of fixed and embedded tissue blocks. Intense labeling was obtained on sections from newly processed as well as 20-year-old tissue blocks.

Protein A–gold is an indirect two-step immunocytochemical technique based on the affinity existing between the antibody and its corresponding antigen and the ability of protein A to bind the Fc fragment of immunoglobulin G. Thus, the labeling protocol includes exposure of the tissue sections to two main reagents – the specific antibody and the protein A–gold complex. This will generate the complex antigen–antibody–protein A–gold as depicted in the diagram in Fig. 4.2. Accordingly, the labeling protocol consists of two main incubation steps, one with the antibody and the other with the protein A–gold complex. Additional steps are, however, needed in the labeling protocol to assure specificity of the final result. The labeling technique is illustrated in Fig. 4.3 and is carried out as follows:

1. Tissue sections mounted on nickel grids are incubated by floating them sections down on a drop of PBS containing 1% ovalbumin (or immunoglobulin-free bovine serum albumin) at room temperature for at least 5 minutes.

Figure 4.4. *Washing grids between incubation setps.*
Left. Grids floating sections down on various baths of buffer solution. **Right.** Agitation of grids on magnetic plate to allow better washing.

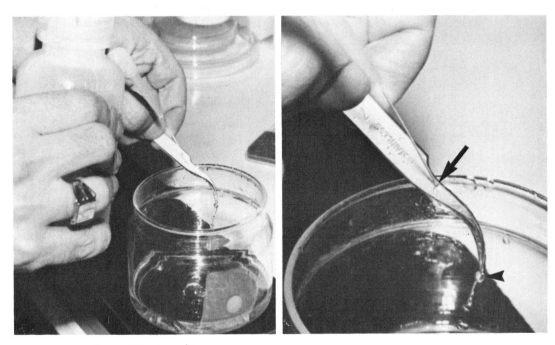

Figure 4.5. *Jet washing.*
Stream of buffer or water solution directed on tweezers (*arrow*). Liquid allowed to run over grid (*arrowhead*).

2. The grids are then transferred (without shaking) to a drop of the diluted antibody. This incubation can be carried out for 2 hours at room temperature or for longer periods of time (up to 24 hours) at room temperature or 4°C. If incubation is carried out at 37°C, it should not last more than 30 minutes.

3. After removal from the antibody solution, the grids are rinsed with PBS in order to remove any antibody molecule that adheres to the sections without binding to its specific determinant. This is performed by transferring the grids, sections down, to small wells of porcelain dishes containing about 1 mL of PBS, and floating them successively on five of these baths for 3 to 5 minutes each. In order to improve the rinsing process, the grids could be swirled by placing the porcelain dishes on a magnetic plate as illustrated in Fig. 4.4 (right).

4. The grids are then incubated on a drop of PBS containing 1% ovalbumin for at least 5 minutes at room temperature.

5. Without rinsing, the grids are transferred to a drop of the protein A–gold complex and incubated for 30 minutes at room temperature. NOTE: It is important that all incubations should be carried out by floating. It is imperative that the reagents have no access to the other face of the grid in order to avoid excessive background due to the adhesion of the reagents to the supporting film.

6. The grids are then thoroughly rinsed to remove unbound protein A–gold complexes. This is carried out by swirling them on various baths of PBS and by jet washing with PBS. At this step the grids can be immersed in the rinsing solutions. Jet washing should be carried out with some precautions: care should be taken not to direct the jet on the tissue sections but rather on the tweezers as illustrated in Fig. 4.5. The stream of washing solution should run over the sections and not underneath.

7. The grids should be rinsed with distilled water and dried. When drying by laying the grids on a filter paper, care should be taken to place them sections up. It is quite important not to dry the sections throughout the procedure until rinsing with the distilled water is done.

8. Tissue sections can be stained according to routine techniques using uranyl acetate and lead citrate before examination with the electron microscope. However, staining protocol can vary, depending on the resin in which the tissues were embedded. For example, with Lowicryl K4M-embedded tissue, staining with aqueous uranyl acetate should be reduced to 4–5 minutes and that with lead citrate to 60 seconds.

In most cases when the tissue has been postfixed with osmium tetroxide, the sections should be pretreated with an oxidizing agent in order to unmask the antigenic sites. We have found that sodium metaperiodate gives optimal results since it has minimal etching effect on the surface of the tissue section (Bendayan & Zollinger, 1983). For this procedure, thin sections of tissues that have been fixed in glutaraldehyde, postfixed in OsO_4 and embedded in Epon are mounted on nickel grids and first incubated on a drop of a saturated aqueous solution of sodium metaperiodate for 30 to 60 minutes at room temperature. After incubation, the grids are rinsed by floating them on several (up to five) baths of distilled water in order to remove any trace of periodate and then transferred to the drop of PBS containing 1% ovalbumin as mentioned in step 1 of the protocol. It is important to keep the periodate solution always at room temperature in order to maintain its degree of concentration at saturation. Treatment of non-osmicated tissue sections with sodium metaperiodate seems to enhance the intensity of the labeling.

Comments on Results Using Immunogold

Large variations exist among antigens concerning their reactivity toward their antibodies after fixation and embedding. These variations, which extend from antigens that are very resistant yielding high intensities of labeling to extreme sensitivity with total absence of labeling, depend on two parameters: the initial amount of antigen present in the tissue and its sensitivity to the chemicals used for tissue preparation. Indeed, since immunocytochemistry is a threshold technique, if the initial number of antigen molecules in the tissue is low, the number that will be exposed at the surface of the section and available for labeling will be beneath the sensitivity of the approach. In addition to the routine techniques that include fixation with glutaraldehyde and osmium tetroxide followed by embedding in epoxy resins, several other protocols have been designed for successful immunocytochemical labeling of sensitive antigens for electron microscopy. The most frequently employed techniques consist of fixation with paraformaldehyde or paraformaldehyde–lysine–periodate solutions, followed by embedding in Lowicryl, LR resins, or glycolmethacrylate resins. We should stress, however, that no rule exists concerning sensitivity of the antigens and, in each particular case, it is advised to perform several tests in order to determine the optimal conditions of labeling (Bendayan et al., 1987). In spite of this, as deduced from experience and from literature documentation, certain generalities exist. Indeed, the antigenicity of secretory peptides and proteins such as those present in endocrine (Fig. 4.6), acinar (Fig. 4.7), and other epithelial cells, usually survives fixation with glutaraldehyde and osmium and embedding in epoxy resins. Some mitochondria (Fig. 4.8) and peroxisomal enzymes (Fig. 4.9) have also been revealed under these conditions of fixation and embedding. This is particularly convenient since it allows for superior ultrastructural preservation with labeling of very high resolution. Other cellular enzymes that probably exist in much lower amounts have been successfully detected in tissue processed without osmium fixation and embeded in Lowicryl or LR resins; they are not labeled when embedded in epoxy resins. Cytosolic proteins, including those from the cytoskeleton (Fig. 4.10), and nuclear, proteins (Fig. 4.11), also require mild fixation and embedding in Lowicryl or LR resins. For extracellular matrix proteins (Fig. 4.12), such as the collagen family of proteins, laminin, and proteoglycans, fixation with glutaraldehyde should be avoided. Only paraformaldehyde or paraformaldehyde–lysine–periodate fixatives are acceptable, followed by embedding in Lowicryl or LR resins. In general, integral membrane proteins are difficult to detect by postembedding immunogold techniques since their levels are relatively low, and they are quite sensitive to fixation. Furthermore, only those molecules exposed at the surface of the tissue section can be detected, which is an additional limitation to their revelation by postembedding immunogold techniques. In such cases, embedding in glycolmethacrylates or even processing by cryoultramicrotomy can be very helpful.

Figure 4.6. (Figure on p. 81) Localization of prolactin in electron micrograph of secretory cell from human pituitary adenoma. Labeling is present in endoplasmic reticulum, Golgi apparatus (ga), secretory granules (g), and mitochondria (m) (glutaraldehyde/osmium/Epon–sodium metaperiodate/anti-prolactin/protein A–gold). Magnification: 65,000 × .

Figure 4.7. (Figure on p. 81) Localization of salivary amylase in electron micrograph of parotid gland from diabetic rat. Labeling mainly present in secretory granules (g), some of which show unlabeled areas. Also visible are mitochondria (m), and acinar lumen (l) (glutaraldehyde/Epon–anti-amylase/protein A–gold). Magnification: 30,000 × .

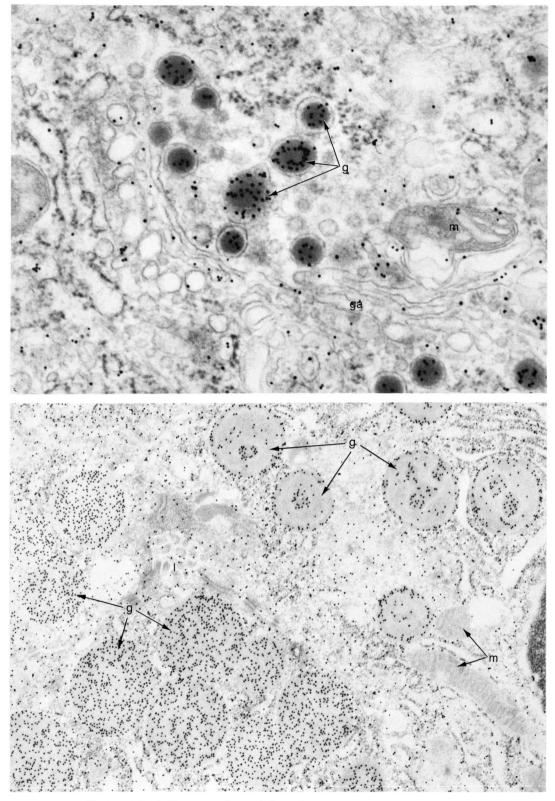

Figure 4.6 (top). Figure 4.7 (bottom). Legend on p. 80

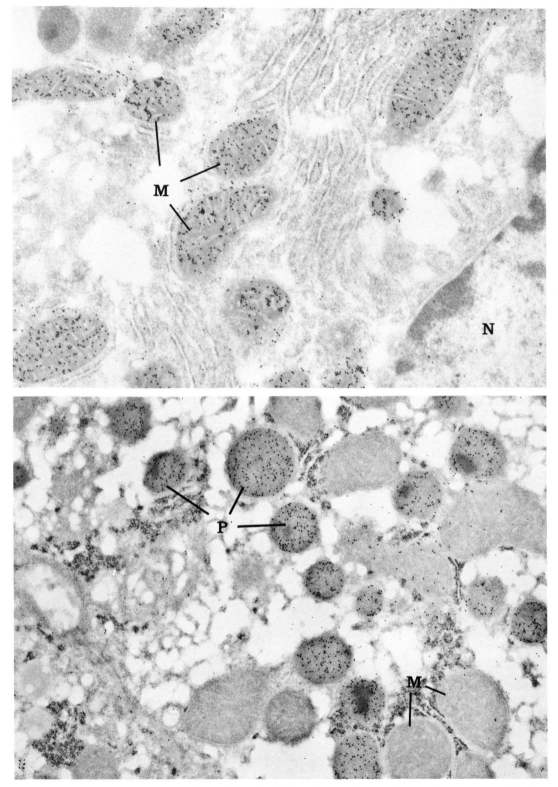

Figure 4.8 (top).   Localization of carbamyl phosphate synthetase (CPS) in electron micrograph of rat liver tissue. CPS labeling restricted to the mitochondria (m). Nucleus (n) also visible (glutaraldehyde/K4M-antibody/protein A–gold). Magnification: 25,000 ×.

Figure 4.9 (bottom).   Localization of catalase in electron micrograph of rat liver. Labeling present in peroxisomes (p). Mitochondria (m) also visible. (Procedures same as Fig. 4.8.) Magnification: 20,000 ×.

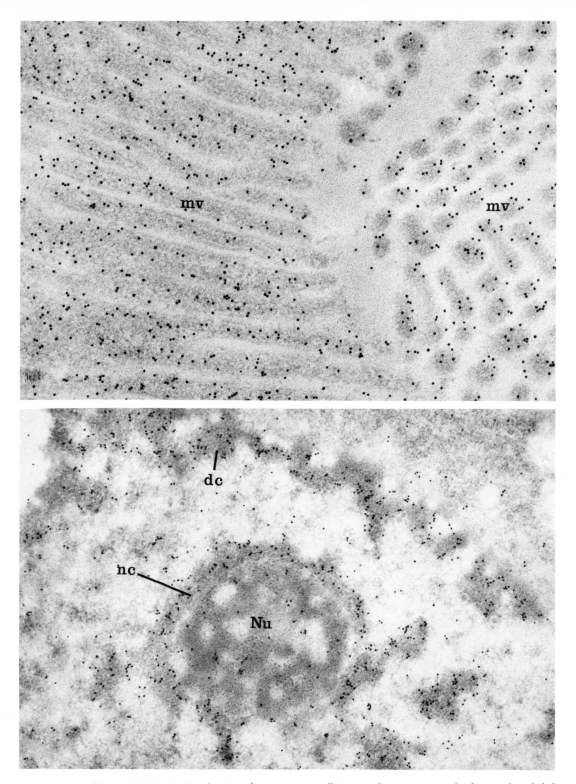

Figure 4.10 (top). Localization of actin on microvilli (mv) in electron micrograph of rat renal epithelial cells. Labeling mainly present in core of microvilli although some particles are visible in association with membranes (glutaraldehyde/GMA–anti-actin/protein A–gold). Magnification: 55,000 ×.

Figure 4.11 (bottom). Localization of nuclear antigen (PSL) in dense chromatin (dc) and nucleolar chromatin (nc) in electron micrograph of rat hepatocytes. Nucleous (nu) also visible (glutaraldehyde/K4M–anti-PSL/protein A–gold). Magnification: 30,000 ×.

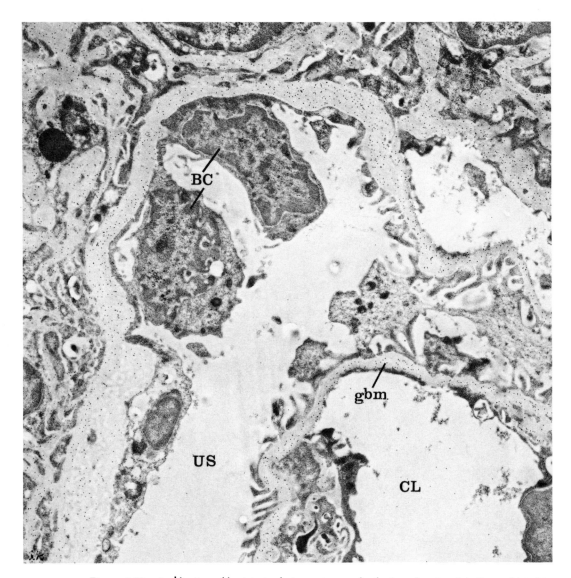

Figure 4.12. Localization of laminin in electron micrograph of rat renal cortex. Labeling restricted to basement membranes, glomerular basement membrane (gbm) as well as that of Bowman capsule (BC). Also visible are capillary lumen (CL) and urinary space (US) (paraformaldehyde/K4M—anti-laminin/protein A—gold). Magnification: 18,000 ×.

Control of Specificity

The specificity of the labeling should be assessed by control experiments. Some of these experiments are of prime importance, while others can be considered as useful but nonessential. The three most important methods are:

1. Incubation of the tissue section directly with the protein A—gold probe, omitting the antibody step. This incubation should take place for 30 minutes at room temperature, similar to the condition used for the labeling protocol. This control experiment can be used to assess the nonspecific adsorption of the probe to the tissue sections.

This adsorption is, in general, negligible but varies from one resin to another and certainly with the degree of polymerization of the resin. For these reasons, this control procedure should be performed each time a new embedding experiment is carried out.

2. Incubation of the tissue sections with an antibody solution to which an excess of the corresponding antigen has been previously added. The incubation is carried out following the same protocol as for the specific labeling, followed by the protein A—gold probe. This control experiment verifies the specificity of the interaction occurring between the antibody and the corresponding antigen. The antibody molecules are competing between the antigenic sites present in solution and those present at the surface of the tissue sections. It should yield very low levels of labeling. The adsorption of the antibody with its antigen should be done either 24 hours prior to labeling and kept at 4°C or a few hours before labeling and kept at room temperature.

3. Incubation of the tissue sections with the antibody, followed by native protein A for 30 minutes at room temperature, and then followed, in a third step, by an incubation with the protein A—gold complex for 30 minutes at room temperature. This control experiment verifies the specificity of the interaction occurring between the protein A—gold and the Fc fragment of the immunoglobulins.

The native protein A should bind to all the Fc fragments present at the surface of tissue sections to prevent any labeling with protein A—gold.

To these control experiments we can add the following procedure that will further confirm the specificity of the labeling obtained.

1. Adsorption of the protein A—gold with an excess of IgG prior to its use for labeling. Competition between IgG in solution and that at the surface of the sections should lead to very low levels of labeling.

2. Incubation of the tissue sections with normal serum in place of the specific one, followed by protein A—gold. Only background staining should be obtained.

3. Incubation of sections of different tissues, known not to contain the antigen under study, with the specific antibody, followed by the protein A—gold. This will permit good evaluation of the nonspecific adsorption of the antibody and the protein A—gold to tissues processed in a similar way as the one displaying the particular antigen.

Additional Recommendations and Problems of Labeling

1. The specificity of the results is totally dependent on the quality of the antibody and on its affinity to its antigen. The antibody should be well characterized and of high titer. Whole antiserum, an IgG fraction, or an affinity-purified antibody can be used with the protein A—gold technique. For monoclonal antibodies, either the supernatant of ascites fluid or purified forms can be used with protein G—gold or protein A/G—gold complexes, although protein A—gold has also been found to react with some monoclonal antibodies. Fab fragments of immunoglobulins should be absolutely avoided because they lack their Fc fragments, the site of interaction with the proteins A, G or A/G. The antibodies should be of the IgG class rather than IgM or IgA because only the Fc portion of IgG will bind Protein A.

2. When protein A–, G–, or A/G–gold complexes are to be used, one has to identify the affinity of these reagents with IgG molecules of different mammalian species. Protein A/G–gold complex displays the highest versatility.

3. Antibodies should be kept frozen until use. They should be frozen as small aliquots to avoid thawing the entire stock of the antibody. Successive freeze–thawing will damage the antibody characteristics.

4. Dilution of the antibody is carried out in PBS, which may contain 0.1% sodium azide. The solution should be kept at 4°C.

5. The extent of the dilution varies from one antibody to another, depending on its titer. For labeling purposes, dilutions and incubation conditions (time and temperature) should be worked out in order to determine optimal requirements (i.e., those yielding the highest specific labeling with an acceptable level of background).

6. Background will occur in all cases. However, it should be kept as low as possible. The reasons for background are numerous. Some can be identified and thus corrected; others are less perceivable.

   a. Specificity of the antibody is most important. Nonspecific adhesion of the antibody molecules to the tissue sections occurs, however, but it can be reduced by incubation of the tissue sections with 1% ovalbumin PBS solution. Prior to the antibody and protein A–gold procedures, incubation in the albumin solution can be allowed to proceed at room temperature for up to 1 hour. Albumin can be added to the rinsing and even to the antibody solutions. In some cases, addition of Tween 20 (0.5%) has been found to improve the results. An additional incubation step can also be performed with 0.15 mol/L glycine for 30 to 60 minutes prior to the albumin step at the beginning of the procedure.

   b. Preincubation with normal serum should be avoided when protein A–, G–, or A/G–gold complexes are used since the IgG molecules present in normal serum will be revealed by the probe.

   c. The length and temperature of incubation with the antibody should be adjusted in accordance with the antibody concentration. In general, using highly diluted antibodies (1/1,000 and higher) with incubation times of 24 hours at 4°C improves the results. Increased dilutions of the antibody will improve the signal by eliminating contaminant low-titer antibodies.

   d. Highly concentrated antibodies will generate high background and also will cause dense deposits that correspond to aggregation of immunoglobulins. These will react with the tissue sections and will be subsequently labeled by the gold reagents [Figs. 4.13 (top left) & 4.14].

Figure 4.13. *Illustrations of various problems that occur during labeling.*
   Top left. Spontaneous aggregation of gold leading to presence of large deposits with specific labeling: peroxisomes (p), mitochondria (m). Magnification: 30,000 ×. **Top right.** Nonspecific adhesion of gold on supporting film (sf) but not on tissue section. Magnification: 60,000 ×. **Bottom left.** Problems of washing. Stream of washing solution has removed gold particles from one area and retained large amounts of background in the other. Magnification: 28,000 ×. **Bottom right.** Presence of major fold (*arrow*) in tissue section, which adsorbs gold probe on one side and prevents labeling on the other. Visible are elastic laminae of a blood vessel (el) and smooth muscle cell (sm). Magnification: 12,000 ×.

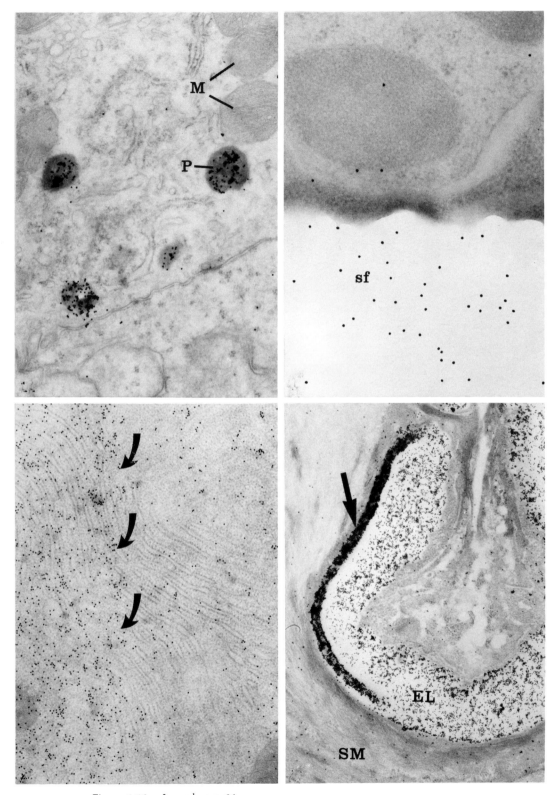

Figure 4.13.   Legend on p. 86.

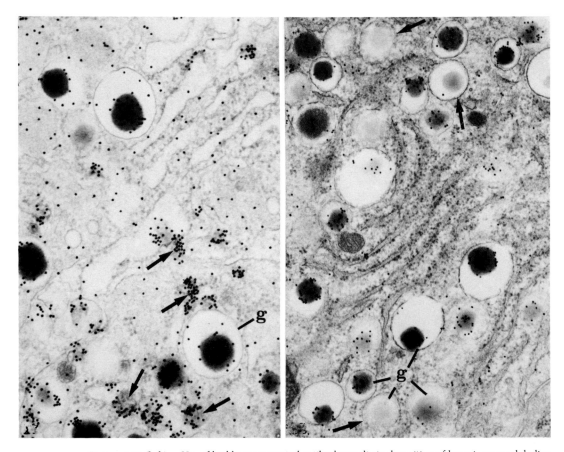

Figure 4.14 (left).   Use of highly concentrated antibody results in deposition of large immunoglobulins complexes (*arrows*), which are labeled by gold particles; secretory granule (g). Magnification: 40,000 × .

Figure 4.15 (right).   Illustration of limits of postembedding immunogold technique. Localization of insulin in a pancreatic B cell of the rat. Labeling is present in most of the secretory granules (g). However, some granules (*arrows*) remain unlabeled since they do not reach the surface of the section. Due to the small amount of material present in the section, these granules appear generally more electron lucent. Magnification: 30,000 × .

e. Dilution of the antibodies should be carried out at least 24 hours prior to being used for labeling. Freshly diluted antibody solutions carry aggregates that may be visualized by labeling (Fig. 4.14).

f. All incubations should be performed in a well-humidified atmosphere in order to avoid evaporation of the solutions and adhesion of the reagents to the sections. This is particularly important for incubations taking place for long periods of time or at 37°C. To humidify the incubation chamber, a soaked filter paper should be placed close to the drops of reagents (Fig. 4.3).

g. Incubations should be carried out by floating the grids on the different solutions (Fig. 4.3), sections face down. It is important not to expose the other face of the grid to the various solutions since the reagents have a tendency to adhere to the supporting films, thus generating high background. Differences between nonspecific adhesion of gold particles on tissue section and supporting film are illustrated in Fig. 4.13 (top right).

h. It is also important to avoid drying the tissue sections during the procedure until the last rinsing with distilled water is performed.

i. If the antibody was raised with a hapten, the antiserum should be blocked by adding an excess of the carrier protein to the antibody solution. This will eliminate eventual cross-reactivities.

j. Ultrastructural preservation and state of the tissue are important. Indeed, necrotic cells and cellular debris have a tendency to adsorb serum proteins and protein–gold complexes.

k. The washing procedure is essential for good labeling. The first washing step should remove the IgG molecules adhering nonspecifically to the sections while retaining those bound specifically to their antigens. The second washing step after incubation with the protein A–gold complex should remove those gold particles that are not bound to the immunoglobulins. Jet washing should be carried out gently but with sufficient pressure. The jet should be directed on the tweezers and not on the sections, letting the stream of buffer run over the sections (Fig. 4.5). Strong jets directed on the sections might remove some gold particles from certain regions while retaining others, generating stripes of nonlabeled areas. Also, they could displace the label and cause the formation of wave artifacts [Fig. 4.13 (bottom left)].

l. The quality of the embedding as well as of the sectioning is important. Poorly infiltrated tissue blocks will generate holes where the gold label will accumulate. Poorly polymerized resins will become sticky and adsorb the gold particles. Folds and vibration artifacts will interfere with the washing procedure, generating strands of gold particles [Fig. 4.13 (bottom right)].

m. Although protein–gold complexes are extremely stable, some aggregation can occur with time. The aggregates correspond to several gold particles closely touching each other. These can be removed by a simple low-speed (2,000 rpm) centrifugation of the protein–gold complex prior to its use for labeling [Fig. 4.13 (top left)].

n. The use of nickel or gold instead of copper grids is an important factor since copper will easily oxidize during the various incubations and contaminate the tissue sections.

Limitations

All postembedding immunogold techniques are faced with a limitation due to the fact that the gold probe does not penetrate the thickness of the section. Only those antigenic sites exposed at the surface of the section by the cutting procedure can be revealed. Structures that are present in a tissue section but not exposed at its surface, while being visible in the microscope, are however unlabeled by the gold probe (Fig. 4.15). However, this limitation has a beneficial counterpart. Since labeling is restricted to the surface, quantitation and comparative evaluations of intensities of labeling are possible.

PROTEIN A–GOLD VERSUS PROTEIN G–GOLD AND PROTEIN A/G–GOLD COMPLEXES

Due to its high binding affinity to the Fc fragment of immunoglobulins G, protein A has become a widely used reagent in various immunoassays (Langone, 1982). Tagged with colloidal gold, the protein A–gold complex has proven to be a specific tool in immunocytochemistry in light and electron microscopy (Bendayan, 1984, 1989). However, while the affinity for immunoglobulin G of some mammalian species such as those of rabbit, guinea pig, and human is quite good, protein A shows low affinity for the immunoglobulins of sheep, rat, mouse, and goat. Alternative reagents are now available. Protein G displays properties similar to protein A, with the advantage of having higher affinities for the immunoglobulins of mouse, rat, goat, and sheep (Åkerström & Björck, 1986). Thus, protein G once tagged to colloidal gold particles was found to be superior to protein A–gold, particularly when working with mouse monoclonal antibodies (Bendayan & Garzon, 1988). Further along this line, recombinant protein A/G has recently been prepared (Eliasson et al., 1988). It displays the combined properties of protein A and protein G with high affinities toward a large spectrum of immunoglobulins. The protein A/G–gold complex has been found to be a very reliable and versatile immunoprobe, which reacts with polyclonal as well as monoclonal antibodies with very high affinity (Ghitescu et al., 1991).

CONCLUSION

In concluding I would like to emphasize the fact that the conditions for the retention of antigenicity after fixation and embedding should not be taken as strict rules. Rather, I recommend that for each new antigen optimal conditions for labeling should be determined. The quality of the results depends mainly on the specificity of the antibody used. The second probe, either the IgG–, protein A–, protein G–, or protein A/G–gold, will only reveal the antigen–antibody complexes formed at the surface of the tissue section. Little background and nonspecific adsorption of the gold to the section occurs in the absence of immunoglobulins. The advantages of the postembedding approach are many. Because the labeling occurs on the surface of the tissue section, the technique does not have the inherent problems of accessibility of the antigen and diffusion of the reagents. The ultrastructural preservation can be excellent since no permeabilization of membrane leading to extraction of cytosolic elements is required. Furthermore, as detailed in other reviews (Bendayan, 1984, 1989), the technique is appropriate for (1) quantitative evaluations of intensities of labeling, (2) multiple simultaneous labeling of various antigens on the same section, (3) combining the gold labeling with various other staining techniques on the same tissue section, (4) study of the dynamics of protein movements inside tissues and cells, (5) amplification of the labeling when the initial signal is low, and (6) labeling at the light microscope level. The only restriction that is common to all immunocytochemical techniques resides in the alteration of the antigens by the tissue-preparation protocols. This can, however, be overcome by the availability of several alternative techniques of fixation and embedding that will preserve antigenicity while providing acceptable ultrastructural details. Since immunocytochemistry is based on the specific labeling of a particular antigen on identifiable cellular structures, the approach described amply fulfills this objective.

REFERENCES

Åkerström, B., & Björck, L. (1986). A physicochemical study of protein G, a molecule with unique immunoglobulin G-binding properties. *J. Biol. Chem.* **261**: 10240–10247.

Baschong, W., & Roth, J. (1985). Lyophilization of protein-gold complexes. *Histochem. J.* **17**: 1147–1153.

Bendayan, M. (1984). Protein A–gold electron microscopic immunocytochemistry: Methods, applications and limitations. *J. Electron. Microsc. Tech.* **1**: 243–270.

Bendayan, M. (1989). Protein A–gold and protein G–gold postembedding immunoelectron microscopy. In *Colloidal Gold: Principles, Methods and Applications*, ed. M. A. Hayat. San Diego: Academic Press, Vol. 1, pp. 33–94.

Bendayan, M., & Garzon, S. (1988). Protein G–gold complex: Comparative evaluation with protein A–gold for high-resolution immunocytochemistry. *J. Histochem. Cytochem.* **36**: 597–607.

Bendayan, M., Nanci, A., & Kan, F. W. K. (1987). Effect of tissue processing on colloidal gold cytochemistry. *J. Histochem. Cytochem.* **35**: 983–996.

Bendayan, M., & Zollinger, M. (1983). Ultrastructural localization of antigenic sites on osmium-fixed tissues applying the protein A–gold technique. *J. Histochem. Cytochem.* **31**: 101–109.

De Mey, J. (1983). Colloidal gold probes in immunocytochemistry. In *Immunocytochemistry: Practical Applications in Pathology and Biology*, eds. J. M. Polak & A. G. E. Pearse. Bristol: John Wright & Sons Ltd., pp. 82–111.

Eliasson, M., Olsson, A., Palmcrantz, E., Wiberg, K., Inganäs, M., Guss, B., Lindberg, M., & Uhlén, M. (1988). Chimeric IgG-binding receptors engineered from staphylococcal protein A and streptococcal protein G. *J. Biol. Chem.* **263**: 4323–4327.

Faulk, W. P., & Taylor, G. M. (1971). An immunocolloid method for the electron microscope. *Immunochemistry* **8**: 1081–1083.

Frens, G. (1973). Controlled nucleation for the regulation of the particle size in monodisperse gold solutions. *Nature* [Phys. Sci.]. **241**: 20–22.

Ghitescu, L., & Bendayan, M. (1990). Immunolabeling efficiency of protein A–gold complexes. *J. Histochem. Cytochem.* **38**: 1523–1530.

Ghitescu, L., Galis, Z., & Bendayan, M. (1991). Protein AG–gold complex: An alternative probe in immunocytochemistry. *J. Histochem. Cytochem.* **39**: 1057–1065.

Handley, D. A. (1989). The development and application of colloidal gold as a microscopic probe. In *Colloidal Gold: Principles, Methods and Applications*, ed. M. A. Hayat. San Diego: Academic Press, Vol. 1, pp. 1–12.

Hobot, J. A. (1989). Lowicryls and low-temperature embedding for colloidal gold methods. In *Colloidal Gold: Principles, Methods and Applications*, ed. M. A. Hayat. San Diego: Academic Press, Vol. 2, pp. 76–111.

Kellenberger, E., & Hayat, M. A. (1991). Some basic concepts for the choice of methods. In *Colloidal Gold: Principles, Methods and Applications*, ed. M. A. Hayat. San Diego: Academic Press, Vol. 3, pp. 1–30.

Langone, J. J. (1982). Protein A of *Staphylococcus aureus* and related immunoglobulin receptors produced by Streptococci and Pneumococci. *Advances in Immunology* **32**: 157–252.

Newman, G. R. (1989). LR White embedding for colloidal gold methods. In *Colloidal Gold: Principles, Methods and Applications*, ed. M. A. Hayat. San Diego: Academic Press, Vol. 2, pp. 48–75.

Romano, E. L., & Romano, M. (1977). Staphylococcal protein A bound to colloidal gold: a useful reagent to label antigen-antibody sites in electron microscopy. *Immunochemistry* **14**: 711–715.

Roth, J. (1982). The protein A–gold (pAg) technique. Qualitative and quantitative approach for antigen localization on thin sections. In *Techniques in Immunocytochemistry*, eds. G. R. Bullock & P. Petrusz. London: Academic Press, pp. 107–133.

Roth, J., Bendayan, M., & Orci, L. (1978). Ultrastructural localization of intracellular antigens by the use of protein A–gold complex. *J. Histochem. Cytochem.* **26**: 1074–1081.

Slot, J., & Geuze, H. (1981). Sizing of protein A–gold probes for immunoelectron microscopy. *J. Cell Biol.* **90**: 533–536.

Tokoyasu, K. T. (1983). Present state of immunocryoultramicrotomy. *J. Histochem. Cytochem.* **31**: 164–167.

Van Bergen en Henegouwen, P. M. P. (1989). Immunogold labeling of ultrathin cryosections. In *Colloidal Gold: Principles, Methods and Applications*, ed. M. A. Hayat. San Diego: Academic Press, Vol. 1. pp. 192–216.

Villiger, W. (1991). Lowicryl resins. In *Colloidal Gold: Principles, Methods and Applications*, ed. M. A. Hayat. San Diego: Academic Press, Vol. 3, pp. 59–73.

5

• *Special Methods*

INTRODUCTION

Because the following methods are used frequently in an EM laboratory (especially a clinical laboratory), it was decided to include them in a separate chapter. The list of chemicals and equipment at the beginning of each technique includes only those unique to each method since general supplies are covered in previous chapters, as are methods for preparing solutions.

CULTURED CELLS

Chemicals and Equipment

- Thermanox coverslips.
- Polypropylene medicine cups – 30 mL.

Method

1. Fix coverslips containing cell monolayer in 2.5% glutaraldehyde in 0.1 mol/L cacodylate bufffer for 30 minutes.
2. Transfer each coverslip to a separate polypropylene medicine cup.
3. Rinse with several changes of buffer.
4. Postfix, dehydrate, and infiltrate with resin in the usual manner (see Chapter 2).
5. Place each coverslip into an embedding mold so that the monolayer is perpendicular to the plane of sectioning (for cross sections). If en face sections are required place the coverslip cell-side down on the surface of a blank, polymerized block of resin to which a small drop of fluid resin has been applied.
6. Cure in a 70°C oven for 6–8 hours or overnight.
7. Remove from the oven and allow to cool 20 minutes.
8. Strip off the coverslip (if embedded en face) with a pair of forceps and trim and cut the block.

Notes on Method

- If cells have been grown on glass coverslips or in a polystyrene culture flask, the monolayer must be separated by enzymatic digestion or by scraping with a "rubber policeman," centrifuged, and encapsulated (see Fine Needle Aspiration Biopsies).
- En face sections are difficult to handle but yield far more information than cross sections. Accurate ultramicrotome alignment is critical.
- Cells must not be permitted to dry at any stage of processing. Solutions should be pipetted off carefully, always leaving a residue covering the cells.
- Multiwell culture dishes are not resistant to solvents and resin and, therefore, cannot be used for processing.
- It is important to rinse well in buffer following fixation. Phosphates present in culture media can cause precipitates when the specimen is postfixed in osmium tetroxide.
- Thermonox coverslips can be cut easily if cross sections of cells are required.

PERIPHERAL BLOOD SAMPLES (BUFFY COAT)

Chemicals and Equipment

- Vacutainer tubes — 7 mL; blue top (containing 3.8% trisodium citrate), green top (heparin), or purple top (EDTA).
- Microspatula (see Fig. 2.4A).

Method (Anderson, 1965)

1. Collect blood in a Vacutainer tube. Mix by inversion.
2. Balance the tube and centrifuge at 600 g (approximately 1,000 rpm) for 20 minutes. The blood will separate into three layers; the top layer will contain plasma, the middle layer leukocytes and the bottom layer erythrocytes. Immediately above the leukocytes, there will be a thin layer of platelets (Fig. 5.1).
3. Using a Pasteur pipette, carefully remove all but a thin layer of plasma.
4. With a Pasteur pipette, slowly run buffered glutaraldehyde down the side of the tube without disturbing the buffy coat.
5. Place the tube in a rack and allow to fix overnight at 4°C.
6. Once the buffy coat is fixed, run the tip of a microspatula (similar to Fig. 2.4A, but straight) around the disk so that it will float free of the test tube wall.
7. Empty the contents of the test tube into a small disposable weighing boat (see Fig. 2.4), which is partially filled with 0.1 mol/L cacodylate buffer. Pick up the buffy coat with forceps and, with gentle agitation, rinse off unfixed erythrocytes.
8. Place the buffy coat, erythrocytes down, on dental wax. With a razor blade, slice the disk to capture erythrocytes, leukocytes, and platelets in each slice.
9. Transfer to EM vials containing buffer.
10. Postfix in osmium tetroxide, dehydrate, infiltrate with resin, and flat embed in the lids of BEEM capsules (see Fig. 2.5A).

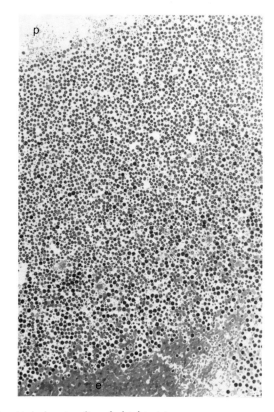

Figure 5.1. Buffy coat showing erythrocytes (e), leukocytes (l), and platelets (p).

Notes on Method

- Sodium citrate is used for the preservation of platelet ultrastructure. Other anti-coagulants (e.g., EDTA) cause depolymerization of platelet microtubules.
- Either heparin or EDTA may be used if the buffy coat is to be prepared within 4 to 5 hours of blood collection. Heparin is best if the blood must be kept from 5 to 24 hours, and EDTA preserves best after 48 hours.

FINE NEEDLE ASPIRATION BIOPSIES

Chemicals and Equipment

- Egg albumin — crystallized, lyophilized (Sigma A5378) or bovine serum albumin — 30% in 0.85% sodium chloride, aseptically filled (Sigma).
- Applicator sticks, pointed.
- Clinical centrifuge.
- Centrifuge tubes, conical — 10 mL.

Solution

Egg Albumin – 20% Buffered

1. Combine 0.2 g egg albumin with 1 mL 0.1 mol/L sodium cacodylate buffer.
2. Store at 4°C. Discard after 1 week.

Method (Mackay et al., 1987)

1. Express aspirate from the syringe into a centrifuge tube containing 2.5% glutar-aldehyde in 0.1 mol/L sodium cacodylate buffer equal to twice the volume of aspirate. Mix gently to prevent clotting. Fix for 30 minutes (minimum).
2. Remove any solid pieces; fix for a further hour and then process. If solid pieces are too small to process, centrifuge at 1,500 rpm for 5 minutes to form a pellet.
3. Remove the supernatant.
4. Dispense a very small drop of 20% egg albumin in buffer (or BSA) onto a piece of dental wax. Transfer the cells with a Pasteur pipette to the albumin and mix with a pointed applicator stick. After 10 minutes, the albumin will gel. It is important to use a small volume of albumin to ensure high concentration of cells within the matrix.
5. Refix and store in 2.5% glutaraldehyde until ready to process.
6. With a razor blade, cut the gel into 1-mm^3 pieces, discarding any albumin that does not contain cells.
7. Wash in buffer, postfix, dehydrate, infiltrate, and embed in the usual manner (see Chapter 2).

Notes on Method

- If the cell/egg albumin mixture fails to solidify, either more concentrated glutar-aldehyde or 10% formalin may be added.
- An alternative method of handling fine needle aspirate biopsies is outlined below (agar method).

PROCESSING CELL SUSPENSIONS FOR EM USING AGAR

Chemicals and Equipment

- Thermal stir plate and small stir bar.
- Pyrex beaker – 50 mL.
- Agar (Difco).
- Pasteur pipettes – 22.5 mm (9 in.).

Method

1. Put 25 mL of sterile distilled water in a 50-mL Pyrex beaker.
2. Add a stir bar.

3. Put the beaker on a thermal stir plate at a speed setting of 3.5 and medium heat.
4. Sprinkle 1 g of agar on top of the swirling water.
5. Heat till agar dissolves and solution becomes clear – approximately 10 min.
6. Meanwhile, put one 9-in. pipette per specimen plus one extra for agar in a 70°C oven.
7. Put the agar in the oven when it clears.
8. Pipette off excess buffer from centrifuged cell suspension that has been fixed in glutaraldehyde, washed in buffer, postfixed in osmium tetroxide, and washed again in buffer.
9. Add 0.5–1 mL of hot agar.
10. Stir gently but do not break up large clumps.
11. Draw the mixture into a hot pipette and slowly squeeze onto a cool slide (it should come out as a solid strip – if not, squeezing is too fast).
12. Slice in a pool of buffer (selecting areas of highest cell density) and process in the normal manner.

Notes on Method

- The cell suspension must be centrifuged and resuspended at each stage of processing until trapped in agar.
- Postfixing in osmium tetroxide enables one to see tiny aggregates of cells.

NASAL BRUSHINGS

An assessment of cilia motility is often of diagnostic value, especially in cases of suspected Kartagener's syndrome. Nasal brushing is an easy means of collecting ciliated cells. References for this method are Hayat (1989) and Mizuhara & Futaesaku (1972).

Solution

Glutaraldehyde–Tannic Acid

1. Combine 50 mL 2.5% glutaraldehyde and 0.5 g tannic acid.
2. Add tannic acid immediately prior to use.

Method

1. Go to the site of a nasal brushing equipped with a light microscope, slides, coverslips, a 1-mL syringe (with 22-gauge needle) filled with room-temperature phosphate buffered saline (PBS), EM vials, and iris forceps.
2. Touch a microscope slide firmly with the nasal brush before putting the brush into a vial of fixative. Enough cells will adhere to the slide for viewing.
3. Apply coverslip with a very small amount of electrolyte solution.
4. By stopping down the iris diaphragm on the microscope, the cells can be seen clearly and an assessment made as to their motility. If they are motile, further procedures might be unnecessary and the following steps can be eliminated.

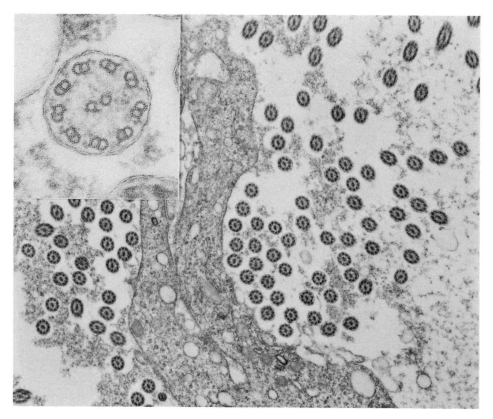

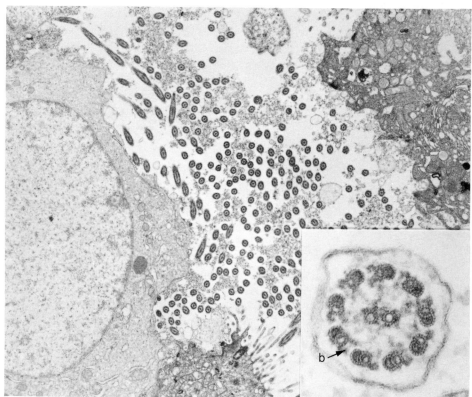

5. Fix the brushing in glutaraldehyde–tannic acid for 60 minutes.
6. Gently tease the pieces of tissue from the brush and wash in 0.1 mol/L cacodylate buffer for 10 minutes. If pieces are too small to process, entrap in egg albumin, BSA, or agar as described above.
7. Postfix in 1% aqueous uranyl acetate for 60 minutes.
8. Do not osmicate or en bloc stain as in Chapter 2 but otherwise process for EM in the normal manner [Fig. 5.2 (top)].

or

Fix in glutaraldehyde–tannic acid, wash in buffer, and process as outlined in Chapter 2, postfixing in osmium and en bloc staining in lead acetate and uranyl nitrate [Fig. 5.2 (bottom)].

PROCESSING TISSUE FOR EM FROM A PARAFFIN BLOCK

Although this produces very poor ultrastructure, it is sometimes the only way a diagnosis can be made. Some structures (e.g., desmosomes, tonofilaments, immune deposits in kidney, viruses) are remarkably well preserved (Fig. 5.3).

Method

1. Identify the area of interest on a stained paraffin section.
2. With a dissecting microscope, find the same area on the paraffin block.
3. Using two pairs of pliers, break a razor blade, to form a sharp point or use a #11 scalpel.
4. Cut around the selected area and then down into the block to produce an inverted pyramid with its apex directly under the area of interest. Since paraffin is brittle, avoid cutting too close to the area of interest and be careful that the piece of tissue does not flip away when lifting it from the paraffin block.
5. Put the tissue into an EM vial and fill the vial with xylene.
6. Deparaffinize in two changes of xylene for 30 minutes each.
7. Change to ethanol–xylene (1:1) for 30 minutes.
8. Hydrate through absolute, 95%, and 70% ethanol for 10 minutes each to 0.1 mol/L cacodylate buffer.
9. Give tissue two changes of buffer, leaving it in the second change overnight.
10. Fix in 2.5% glutaraldehyde in 0.1 mol/L cacodylate buffer for at least 30 minutes.
11. Mince into 0.5 mm^3 and process in the normal manner.

Alternate method, if speed is essential

1. Deparaffinize as above.
2. Give the tissue three 5-minute rinses in absolute ethanol.

Figure 5.2.   (Figure on facing page). *Electron micrographs of nasal brushing specimens.*
   Top. Cross sections of cilia fixed in glutaraldehyde–tannic acid and en bloc stained with uranyl acetate. Grids were contrasted with 3% uranyl acetate in 30% ethanol. Note: Some inner dynein arms appear to be missing. **Bottom.** Cross sections of cilia fixed as above, postfixed in osmium, and en bloc stained as in Chapter 2. Grids were contrasted with uranyl acetate and lead citrate. Both inner and outer dynein arms (d) are visible at high power.

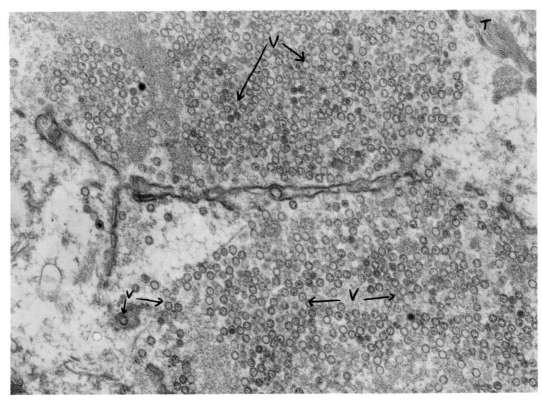

Figure 5.3. Electron micrograph of a punch biopsy of skin lesion from a transplant patient. Vasculitis or scleroderma were suspected. Immunoperoxidase staining for four types of virus was either negative or weakly positive. Paraffin block processed for EM revealed abundant viral structures. Tonofilaments (t), viruses (v).

3. Rinse twice in 99.5 mol. % acetone for 5 minutes each.
4. Infiltrate in grades of resin and cure in the normal manner.

PROCESSING A PARAFFIN SECTION FOR EM

This should be done only if no other tissue is available as ultrastructural preservation is extremely poor. Steps are carried out in a Coplin jar.

Method

1. Remove coverslip with xylene (this can be done in a 60°C oven and will take from 30 minutes to several hours).
2. Remove coverslipping medium with two changes of xylene.
3. Place section in xylene–acetone (1:1) twice for 15 minutes each.
4. Place section in acetone–ethanol (1:1) for 15 minutes.
5. Place section in ethanol (add a pinhead of toluidine blue powder to make the section visible) for 3 minutes.
6. Place section in ethanol–acetone (1:1) for 15 minutes.
7. Rinse section in acetone, twice for 15 minutes each.

8. Infiltrate with 30% resin in acetone for 10 minutes, 60% resin in acetone for 10 minutes, and pure resin for 15 minutes.

9. Fill BEEM capsules to form a convex meniscus.

10. Invert slide over the BEEM capsules making sure the area of interest is covered and the slide is well supported.

11. Polymerize at 70°C overnight.

12. Split the capsules off. Invert a tin of compressed air (e.g., Dust Off Plus) and spray the back of the slide. This will freeze the slide, making it easier to separate the capsules.

13. There will be only 3–5 μm of tissue available, so align the block carefully and either cut thin sections directly or take a maximum 0.25-μm orientation section.

REFERENCES

Anderson, D. R. (1965). A method of preparing peripheral leucocytes for electron microscopy. *Journal of Ultrastructural Research* **13**: 263–268.

Hayat, M. A. (1989). *Principles and Techniques of Electron Microscopy: Biological Applications.* 3rd ed. Boca Raton: CRC Press, pp. 298–303.

Mackay, B., Fanning, T., Bruner, J. M., & Steglich, M. C. (1987). Diagnostic electron microscopy using fine needle aspiration biopsies. *Ultrastructural Pathology* **11**: 659–672.

Mizuhara, V., & Futaesaku, Y. (1972). New fixation for biological membranes using tannic acids. *Acta Histochemica Cytochemica* **5**: 233–236.

6

by Peter Maloney

• *The Electron Microscope*

INTRODUCTION

The purpose of this chapter is to acquaint not only the novice electron microscopist, but also the seasoned user, with the principles of physics and mechanics as they apply to a typical transmission electron microscope (TEM). It is not the intention of the author to make electron-optic design engineers out of beginners but rather to give a basic understanding of the various systems that comprise the modern TEM. This chapter will also discuss certain criteria that must be met in order to optimize a TEM image and how the user can identify and subsequently correct common faults.

Figure 6.1 shows a typical modern TEM, which may be found in any research, quality control, or hospital pathology laboratory. To anyone who has never used a TEM the mere size and number of controls can be quite intimidating. However, if one thinks of an electron microscope as a large light microscope (LM), with which most people are familiar, then the TEM becomes less intimidating.

Figure 6.2 (left) is a ray diagram of a typical compound light microscope. It can be seen that four basic components are required in order to form a magnified image of an object at the eye. (1) A *source of illumination* is needed, which can be as simple as a light bulb. (2) A lens is needed to focus the light onto the specimen; this is called a *condenser lens*. After the light has passed through the specimen, a magnified image is formed by the (3) *objective lens*. This image is further magnified by (4) a *magnifying lens* (which in this case is merely the eyepiece), with the final magnification being the product of the objective and magnifying lens powers.

Figure 6.2 (right) is a ray diagram of a typical TEM and if one simply inverts Fig. 6.2 (left), it can be seen that the same four basic components are required in a TEM to produce a magnified image. (1) An electron gun, which emits a stream of electrons rather than visible light, is now the source of illumination. (2) A condenser lens is still required to focus the illumination onto the specimen (most TEMs use two condenser lenses). (3) An objective lens is still used to form the first image of the

Philips Electronics, Ltd., London, Ontario, Canada.

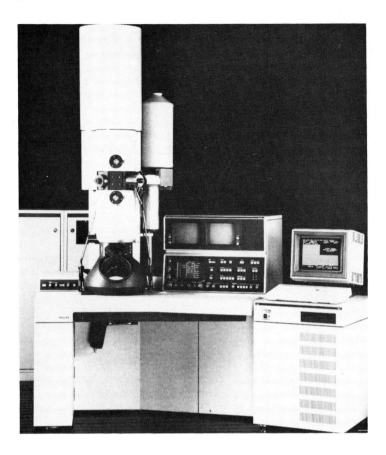

Figure 6.1. Modern transmission electron microscope.

specimen, and a series of (4) magnifying lenses follow to create the final magnified image.

This comparison between a TEM and an LM has been oversimplified but it should be clear now that, although a TEM is much larger and more complex than a simple LM, its functional components remain the same. Figure 6.3 is a cross-section of a typical TEM column. Although it appears much more complex than the simple three-lens example already discussed, the lenses can still be seen as belonging to three functional groups: (1) illumination, (2) image forming and focusing, and (3) final magnification.

When one thinks of an electron microscope and its uses, high magnification is usually thought to be the prime objective. As was shown previously, the final magnification in a compound LM is merely the product of the objective lens and eyepiece powers, so why can we not just use more and more powerful light microscopes to produce infinitely high magnifications? In other words, why do we need electron microscopes if simple light microscopes can produce any power of magnification required? The answer — *resolving power* — brings us to a very important concept in any type of microscopy. The resolving power of a microscope can be defined simply as the ability of the microscope to discriminate between two separate objects. In order to appreciate how and why objects are resolved, it is necessary to understand something about light and how it is transmitted through space.

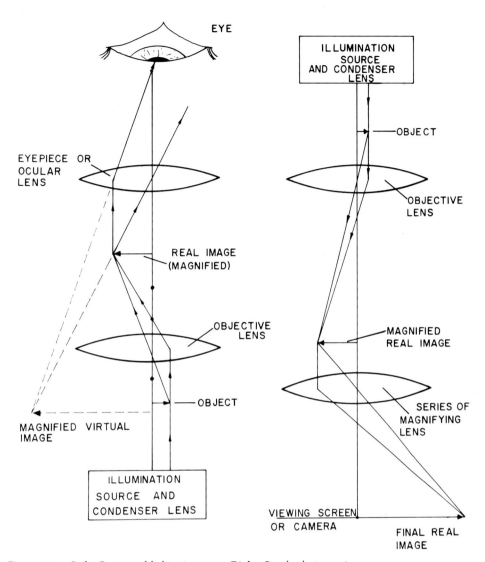

Figure 6.2. **Left.** Compound light microscope. **Right.** Simple electron microscope.

Light can be described as a series of electromagnetic waves propagating through space much as ripples do in a pond. The distance from one wave peak to the next is called the wavelength (lambda, λ) and the visible light spectrum contains light with wavelengths from 0.4 to 0.7 μm. It can be shown that the resolving power of a microscope is limited by the wavelength of light used and, in general, in order to be resolved, objects must be greater than one-half the wavelength of light. Therefore, for the LM the theoretical resolution is 0.5 μm/2 or 0.25 μm. Since most viruses and many intracellular structures are less than 0.25 μm in diameter, an LM will be unable to resolve them in fine detail no matter how many times the image is magnified. If the resolving power of an LM is 0.25 μm, then a maximum magnifying power of 1,000 $\times$ is all that is needed to utilize the resolving capabilities, i.e., 1,000 $\times$ 0.25 μm = 0.25 mm, which can be resolved by the human eye.

How can resolving power be increased? Since resolving power is limited by the

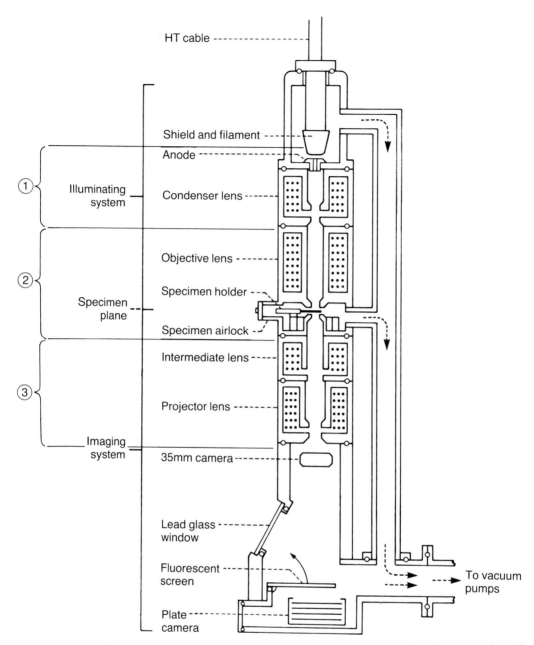

Figure 6.3. Cross section of typical transmission electron microscope. (1) The filament, anode, and condenser lenses forming illumination system. (2) Image forming and focusing by the objective lens. (3) Lenses magnifying image below the objective lens.

wavelength of light or radiation used, then resolution should increase by using radiation of a shorter wavelength. Unfortunately, as wavelength decreases we move past the visible light spectrum and get into higher energy radiation such as ultraviolet (UV) and X-radiation, all of which are harmful to us and present problems when displaying a final image. However, if a beam of electrons is substituted as the shorter wavelength radiation, the resolving power can be greatly increased.

An electron beam does not fit the classical definition of an electromagnetic wave such as visible light. However, it does have a wavelength thousands of times shorter than that of visible light. If we use de Broglie's formula for calculating the wavelength of particles (e.g., photons or electrons), then the wavelength of an electron beam in a TEM with an accelerating voltage of 100,000 volts will be in the order of 0.005 nm or 0.000005 μm. Using radiation of this wavelength, a resolution 100,000 times better than with visible light should be possible but due to design and physical limitations, the resolving power of a typical TEM is in the order of 1,000 times better than an LM. With this resolving power, a TEM can utilize that much more magnification; in fact, magnifications of 1 million times are not uncommon.

In order to focus this type of radiation, glass lenses found in LMs have given way to magnetic lenses, resulting in a much larger and more complex instrument. To further complicate matters, as the wavelength decreases far below that of visible light, energy starts to be absorbed by air molecules, and it is for this reason that the electron-beam path must be maintained under a vacuum.

We have now seen that a functioning TEM requires a source of electrons (called an electron gun), an illumination system, image-forming lenses, magnifying lenses, a means to view and record the final image, and a means to maintain the electron path (or column) under a vacuum. The next part of this chapter will examine each of these components briefly and will include basic information on their working principles and terminology.

VACUUM SYSTEM

Most modern TEMs utilize three stages, or levels, of vacuum, as shown in Fig. 6.4. The path of the electron beam, which includes the electron gun, the lenses, and the specimen, is maintained at *ultra-high vacuum* (UHV) by an *ion-getter pump* (IGP). The lower part of the column, containing the viewing screens and photographic material, is maintained at a lower vacuum, usually referred to as *high vacuum* (HV), by either an *oil diffusion pump* (ODP) or by a *turbo molecular pump* (TMP). This vacuum level is backed by a *rotary vacuum pump*.

The following are some useful relationships concerning vacuums:

$$1\,\text{mmHg} = 1\,\text{torr}$$
$$1\,\text{E}^{-3}\,\text{torr} = 1\,\text{E}^{-3}\,\text{mmHg} = 1\,\text{E}^{-6}\,\text{mHg} = 1\,\mu\text{mHg}$$
$$1\,\text{newton/m}^2 = 7.501\,\text{E}^{-3}\,\text{mmHg} = 1\,\text{E}^{-5}\,\text{bar} = 1\,\text{pascal}$$
$$0.75\,\text{torr} = 1\,\text{mbar} = 100\,\text{pascals}$$

Until recently, vacuum has been expressed using torr or microns; however, it is now more common to see it written in terms of mbar and pascals, with pascals being the preferred unit in the SI system.

Rotary Vacuum Pump

The rotary vacuum pump, sometimes referred to as a pre-vacuum pump (PVP), is responsible for maintaining sufficient backing vacuum for the high-vacuum pump and

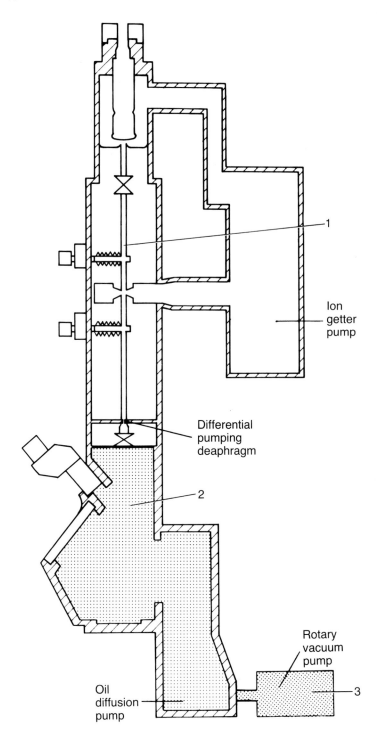

Figure 6.4. Three levels of vacuum found in most transmission electron microscopes. (1) Ultra-high vacuum in beam path. (2) High vacuum in area of specimen stage and camera. (3) High vacuum backed by the rotary pump.

for exhausting any pumped air into the atmosphere. Most rotary pumps are of the "vane" type. Figure 6.5 depicts a typical rotary vacuum pump.

During operation the acentric-mounted rotor is spun, usually via a direct-coupled A.C. motor. As the rotor spins, the vanes sweep along the stator and are compressed, thus creating a smaller volume as they rotate. When air enters through the vacuum

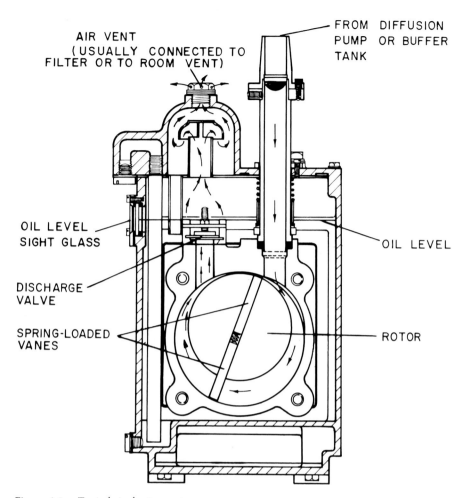

Figure 6.5. Typical single-stage rotary vacuum pump.

connection, it is trapped by the rotating vanes and is subsequently compressed. When this compressed gas is rotated past the one-way exhaust valve, it is expelled into the atmosphere. The entire pumping mechanism is immersed in oil to prevent air leaks and to lubricate moving parts. A pump of this type, with just one rotor, is referred to as a *single-stage pump,* and the ultimate vacuum is limited to approximately 0.1 mbar due to leakage of gas back across the rotor and its seating. This pump can be improved by simply adding another pumping chamber in series with the first one. Now the first chamber backs the vacuum in the second one; this type of pump is known as a *two-stage pump* and can achieve pressures as low as 0.001 mbar or better.

The disadvantage of a rotary pump is that any condensible vapors such as water vapor that enter the pump from the microscope will be compressed to their saturation point, causing liquid to be mixed with the lubricating and sealing oil. This causes deterioration of the ultimate vacuum as well as a breakdown of the lubricating and sealing properties of the oil. The best way to deal with this problem is to ensure that any volume of air to be pumped is free of water vapor. Since some vapor will always be encountered during normal operation (especially moisture from photographic material located in the TEM vacuum), this vapor must be removed before it enters

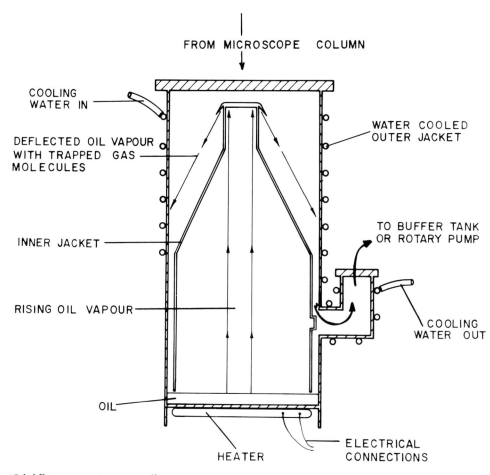

FROM MICROSCOPE COLUMN

COOLING WATER IN

DEFLECTED OIL VAPOUR WITH TRAPPED GAS MOLECULES

INNER JACKET

RISING OIL VAPOUR

OIL

HEATER

WATER COOLED OUTER JACKET

TO BUFFER TANK OR ROTARY PUMP

COOLING WATER OUT

ELECTRICAL CONNECTIONS

Figure 6.6. Oil diffusion pump in cross section.

the pump, or the oil must be exchanged for fresh oil on a regular basis. The most common method of vapor removal involves a *vapor trap* located at the pump inlet. This trap is filled with a material (usually aluminum oxide pellets) that will adsorb water vapor. When the pellets become saturated, they are simply exchanged for new ones.

Another method of keeping the pump oil free of water involves limiting the compression of the pumped gases to a pressure slightly below the saturation point of water vapor, but sufficient for the pump to function. This is called *gas ballast.*

Oil Diffusion Pump

Unlike the rotary pump, the oil diffusion pump (ODP) (Fig. 6.6) contains no moving parts. Instead, it relies on heavy oil vapor streaming up inside the inner jacket, then being deflected downward by one or more jets. These vapor molecules carry with them any gas molecules encountered at the top of the pump, thus creating a lower pressure at the top of the pump than elsewhere in the column. Gas molecules throughout the column are now free to diffuse to this zone of lower pressure and are subsequently pumped away by the oil vapor. After the oil vapor has been deflected, it contacts the side wall of the pump (which is usually cooled by a water jacket) and condenses back into its liquid state. The oil then runs down the side of the pump to the oil reservoir where it is vaporized again by an electric heater. The gas molecules, which enter the pump with the oil vapor, are then pumped away by the backing pump; in this case a rotary pump is used. The ultimate vacuum for an ODP can be as good as $1 E^{-8}$ torr ($1 E^{-6}$ Pa) and is usually measured by a *penning ionization gauge* (Philips gauge).

The advantages of using an ODP include fast pumping and simple, trouble-free operation. The main disadvantage of an ODP is that, if not operated properly, oil vapor can be sucked back (backstream) into the microscope column, thus contaminating ultraclean surfaces. Backstreaming can occur if the temperature of the water jacket that condenses the oil vapor is allowed to exceed the critical level of oil-vapor condensation. When this happens, the vapor is free to diffuse, or backstream, into the column. The usual safeguard for this problem involves placing a temperature sensor on the water jacket which will shut off the pump should the temperature exceed a safe level.

Backstreaming can also occur if the backing pressure to the ODP rises above some critical value, known as the *critical backing pressure* (CBP). This situation can be avoided by monitoring the backing pressure to the ODP and ensuring that it stays below the CBP. Backing pressure is usually measured by use of a *Pirani gauge* located somewhere between the rotary pump and the ODP.

The backing pressure can be maintained at a level below the CBP by the use of a continuously running rotary pump, or a vacuum reservoir (known as a buffer tank) can be added between the rotary pump and ODP. In the latter case, the rotary pump runs intermittently to ensure that the buffer-tank pressure never exceeds the CBP. The advantages of a continuously running rotary pump are its simple and less costly electronics and, since a buffer tank can take up considerable space, a more compact desk design. The main disadvantage of this pump is that it is a constant source of noise and potential vibration, which may cause deterioration in image quality. For this reason, a continuously running pump is usually placed outside the microscope desk unit (i.e., on the floor).

The advantage of a buffer–tank system is that the pump only runs for short periods so that the noise and vibration problems are practically eliminated. Also, the pump is hidden within the desk, thereby making a neater looking instrument. The disadvantage of this system is that some vibration may be noticed on the screen when the pump first starts up. This does not affect photography as the microscope's logic circuits prevent the pump starting until after an exposure is completed.

In cases where backstreaming of oil vapor may be a problem, an alternative high vacuum pump called the turbo molecular pump (TMP) is sometimes used. Simply described, this pump consists of a series of propeller-type blades, which spin at an extremely high speed to cause a net flow of air molecules into the pump. As mentioned, the advantage of TMP over an ODP is the absence of backstreaming oil vapor. Disadvantages include a loud high-pitched sound, high price, and low dependability. These disadvantages, plus the fact that backstreaming in modern ODPs is barely detectable, are the reasons ODPs remain the high vacuum pump most widely used today.

Over the past decade, TEM users have found more and more uses for TEMs that

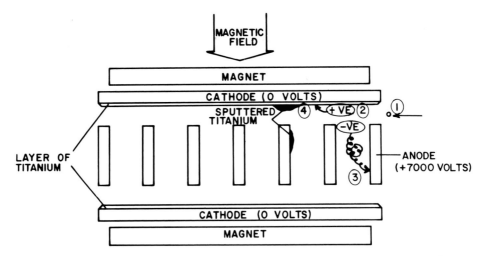

Figure 6.7. Ion getter pump in cross-section. (1) Gas molecules enter pump. (2) Molecules are ionized into +ve and −ve pairs by high voltage between anode and cathode. (3) −ve ions travel toward anode in elongated helical path, thus creating many secondary ion pairs due to collisions with other gas molecules. (4) +ve ions collide with cathode, thus sputtering titanium from site of collision to anode pump body and other areas of cathode. Titanium forms stable compounds with active gas molecules and traps them. Nonreactive gases can be physically trapped on pump surfaces by sputtered titanium.

require vacuums higher and cleaner than can be achieved by using an ODP. Therefore, manufacturers have been forced to adopt a third type of vacuum pump that can create a higher level of vacuum than an ODP, while at the same time introducing no contaminants to the column. The pump used to achieve this ultra-high vacuum is called an ion-getter pump (IGP). Unlike the pumps discussed so far, which evacuate gas molecules to the atmosphere, an IGP captures and stores these molecules. As the name implies, an IGP has two different pumping actions, ion pumping and gettering.

Ion Pump

As gas molecules enter the pump (Fig. 6.7), ionization by the high-voltage field between cathode and anode produce electron and positive ion pairs. The negatively charged electrons are accelerated toward the positively charged anode. As they move toward the anode they collide with other gas molecules, which may lose one or more electrons, thus creating more electron/positive ion pairs. By placing the pump within a strong magnetic field, an electron will move in a helical manner, thus artificially lengthening its path length to the anode. By lengthening the path, the probability of collision and subsequent electron/ion pair production increases, and the ionization efficiency is increased.

Gettering

The positive ions produced by the ionization pump collide with the cathode, dislodging small particles of titanium and "sputtering" them onto the anode, pump body, and opposite cathode. This titanium forms stable oxides and nitrides with active gases such as oxygen, nitrogen, carbon monoxide, carbon dioxide, and water, which in turn are sputtered from the cathode without breaking down.

A second pumping action of gettering is strictly mechanical. Nonreactive gases, such as helium, neon, argon, krypton, and xenon, as well as active gases are buried on the pump surfaces by sputtered material.

Since the active pump material, in this case titanium, will eventually be all sputtered from the cathode, the IGP has a finite lifetime and must be replaced at some point. The usual lifetime is many years.

The electrical current resulting from electrons reaching the anode can be used as a direct indication of the vacuum level within the pump. As the vacuum improves there are less gas molecules to be ionized; therefore, less electrons are liberated and the current to the anode is reduced. This anode current can be presented in an analog format on a vacuum meter or can be converted to a digital signal and displayed on a numeric readout. The ultimate vacuum for a typical IGP is less than $1\,E^{-8}$ torr.

The main problem encountered with IGPs comes when large amounts of water are pumped. Titanium reacts very readily with water to form an oxide and liberate hydrogen, which cannot be quickly adsorbed. For this reason IGPs are usually not turned on until the volume to be pumped has been evacuated to at least $1\,E^{-3}$ torr.

VACUUM MEASUREMENT

Pirani Gauge

As was stated earlier, the Pirani gauge is used to measure low vacuum levels such as those attained by a rotary pump, i.e., down to $1\,E^{-3}$ torr. This gauge consists of a thin, heated tungsten wire exposed to the vacuum. If the vacuum is poor, there will be many gas molecules surrounding the wire, which will conduct heat away. This reduces the temperature and thus the resistance of the wire, which results in a high electrical current through this wire. As the vacuum improves, there will be fewer gas molecules surrounding the heated wire to conduct heat away and thus lower its temperature. As a result, the wire's resistance will rise and electrical current will be reduced. Since the flow of electrical current through the wire is proportional to the level of vacuum in the gauge, it can be calibrated to read the actual vacuum level in any units desired.

Penning Gauge

This gauge is used to measure high vacuum levels up to $1\,E^{-8}$ torr and is similar to the ion pump discussed as part of the IGP. Gas molecules entering the gauge move between two parallel plates, the anode and cathode, which are differentially charged to 2,000 volts D.C. The anode and cathode are placed between the poles of a powerful magnet; therefore, when any gas is ionized, the resulting electrons spiral toward the cathode. This artificially lengthened path results in more collisions with other gas molecules and thus increases ionization current, which can be read by a meter. The current is then calibrated in terms of vacuum and displayed either in an analog or digital format.

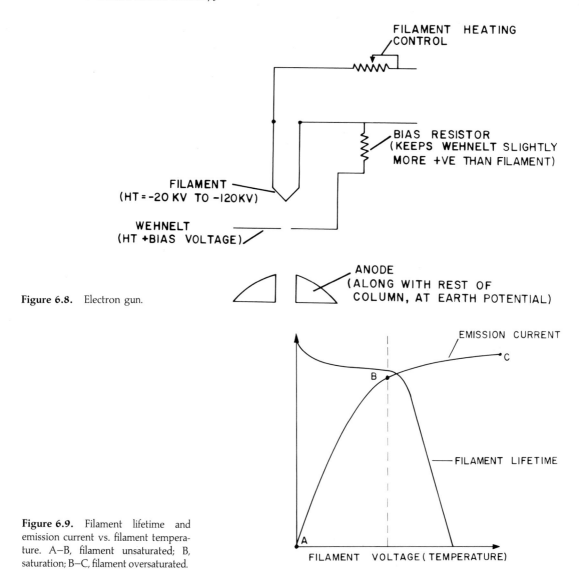

Figure 6.8. Electron gun.

Figure 6.9. Filament lifetime and emission current vs. filament temperature. A–B, filament unsaturated; B, saturation; B–C, filament oversaturated.

MICROSCOPE COLUMN

Electron Gun

The purpose of the gun is to provide a source of electrons for the beam and to accelerate them across a high voltage, down the column, through the specimen, to the viewing screen. A method of varying the amount of electrons transmitted, known as *emission*, must also be provided in order to control image brightness.

The high voltage is generated by circuitry usually submerged in a tank (*high tension tank*) of insulating oil or embedded within a container of epoxy, and is usually adjustable from 20 kV to 120 kV. This high voltage is applied to the gun in such a manner that there is a large potential difference between the filament and anode (Fig. 6.8). A small electrical current is passed through the filament (usually constructed of tungsten), in order to heat the tungsten and stimulate thermonic emission, i.e., a state in which electrons can leave the surface easily due to their increased kinetic energy. The anode, which is 20,000 to 120,000 volts more positive than the filament, accelerates the freed electrons from the filament, through a hole in its center and down the column.

By placing an electrically charged screen, sometimes called a *Wehnelt*, close to the filament and maintaining its voltage slightly less than that of the filament, the total number of electrons accelerated down the column (displayed as the emission current) can be controlled. Since electrons are negatively charged, the -100 to -500 V charge on the Wehnelt acts to repel some electrons coming off the filament and thus reduces the number of electrons available to be accelerated down the column. By changing the Wehnelt voltage (*bias voltage*), the emission current and thus image brightness, can be controlled.

When a tungsten filament is heated, one may think that as the filament gets hotter, more electrons are freed, resulting in higher emission current. This relationship does hold true until a certain filament temperature is reached; then any further increase in temperature results in little, if any, increase in emission (Fig. 6.9). This point is referred to as filament *saturation*. As the filament temperature increases, its lifetime decreases; therefore, if the temperature is set above saturation point, many hours of filament life are needlessly wasted. On the other hand, if filament temperature is set below saturation, an insufficient number of electrons are available, resulting in uneven illumination. Oversaturation of the filament is the most common mistake made by microscopists of all levels of experience. Even a slight oversaturation can reduce filament lifetime considerably; therefore much care should be exercised.

Tungsten filaments are the most common type used today, but in cases where a more intense and coherent beam is required (e.g., for analytical TEM), another type of filament constructed from a single crystal of lanthanum hexaboride (LaB_6) is used. Since LaB_6 filaments can cost up to 100 times more than those made of tungsten, their use is usually limited to those applications where a high-intensity, coherent beam is critical.

Condenser Lenses

The purpose of condenser lenses is to collect accelerated electrons from the gun and focus them onto the specimen with as little loss as possible. Early TEMs had only one condenser lens, which was placed midway between the gun and specimen [Fig. 6.10 (left)]. This system worked fine for focusing electrons to a spot. However, the smallest spot it could produce was only as small as the spot found at the gun crossover (just below the anode). This spot is typically no smaller than 10 μm in diameter, which is therefore the smallest spot that can be focused on the specimen. This means that if the final magnification is 100,000 $\times$, the final spot will have a diameter of 1 meter! Since a typical viewing screen is only 15 cm in diameter, it is easy to see that most of the electrons will miss the screen and be wasted. In this example only about 3% of the electrons striking the specimen actually contribute to an image on the screen, and this figure gets worse as magnification is increased. To maintain screen brightness as magnification increases, the number of electrons within the 10-μm spot must increase. This increase in emission current, as has already been shown, reduces filament life.

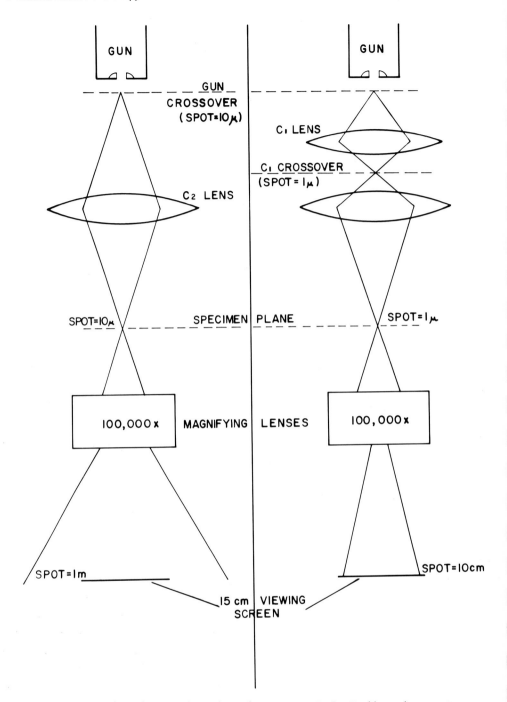

Figure 6.10. Condenser lenses. **Left.** Single condenser system. **Right.** Double condenser system.

Most modern TEMs have overcome this problem by reducing the spot size (prior to the condenser lens) to a value less than that produced by the gun alone, i.e., less than 10 μm. This is done by adding another condenser lens (*C1 lens*) between the gun and the second condenser lens (*C2 lens*) [Fig. 6.10 (right)]. Since the spot area is proportional to the square of the spot radius (area = πr^2), reducing the spot size even

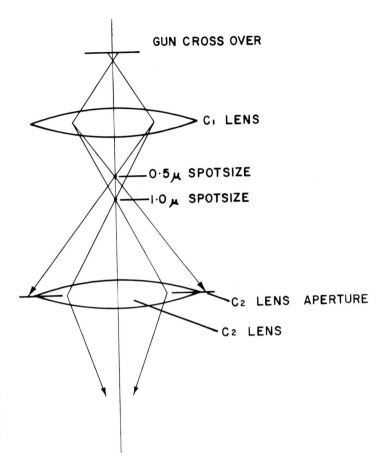

GUN CROSS OVER

C₁ LENS

0·5 μ SPOTSIZE

1·0 μ SPOTSIZE

C₂ LENS APERTURE

C₂ LENS

Figure 6.11. Below a critical spot size (1.0 μm in this example) some electrons cannot pass through the C2 lens aperture and are lost. This results in reduced screen brightness.

by a factor of 2 results in the final spot area at screen level being reduced by a factor of 4, which also means the electron density at the screen is now four times greater. Now, because the screen is brighter, the emission current can be kept lower, which means longer filament life. Another benefit of smaller spot size is that the damaging effects of the beam can be restricted to only that area under observation, thus saving more of the specimen for future use.

In theory, it would be nice to be able to keep the spot size small enough, at any magnification, to transfer all electrons to the screen. In practice it is found that if the C1 lens reduces the spot size too much, many of the resulting electrons fall outside of the C2 aperture and therefore do not contribute to the resulting spot at the specimen (Fig. 6.11), i.e., the electron density of the final spot is reduced. The result is a smaller spot but one that is reduced in brightness. Because of this the C1 lens is usually left at a fixed level, which will result in a final spot with optimal electron density.

The capability of producing spot sizes less than 10 nm is present in high performance analytical TEMs for purposes of microdiffraction and X-ray analysis, where a small spot is more important than final screen illumination.

Now that the spot size has been determined by C1, C2 can be used to vary the intensity of illumination either by under or over focusing. If by focusing C2 the maximum number of electrons will be transferred to the screen, it would seem reasonable to operate C2 in this mode at all times. In practice the use of a focused C2 lens

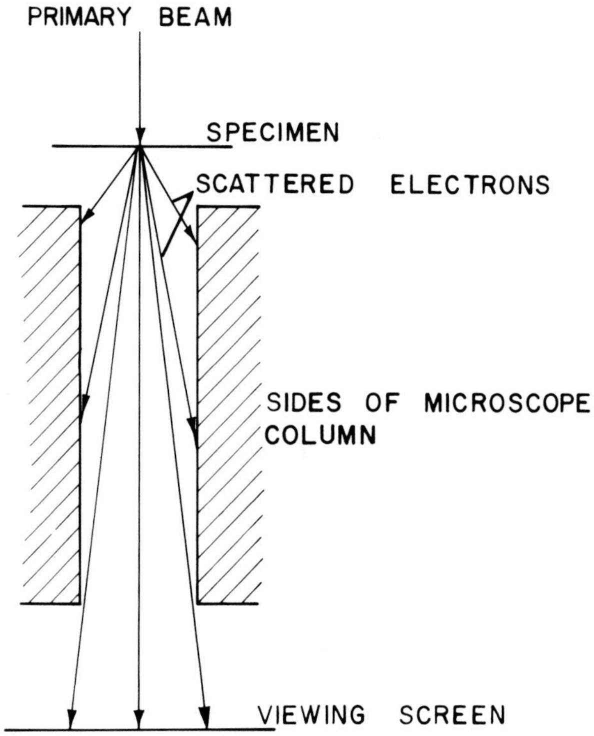

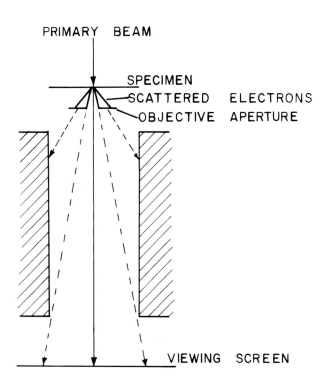

PRIMARY BEAM

SPECIMEN

SCATTERED ELECTRONS

SIDES OF MICROSCOPE COLUMN

VIEWING SCREEN

Figure 6.12. Transmitted scattered electrons. If not intercepted by the sides of the column, these electrons can contribute to the final image.

PRIMARY BEAM

SPECIMEN

SCATTERED ELECTRONS

OBJECTIVE APERTURE

VIEWING SCREEN

Figure 6.13. Objective aperture shown intercepting scattered electrons that would have contributed to final image.

does produce a very bright spot, but the image is blurred and lacks detail. It can also be seen that if the C2 lens is overfocused, the level of illumination drops and resolution improves dramatically (why this happens will be discussed later). Therefore, the C2 lens should be used as far overfocused as possible while still allowing sufficient illumination at the screen.

Objective Lens

The function of the objective lens is to produce the first image of the specimen and focus it in the correct plane (Fig. 6.12) in order that the magnifying lenses can produce the final image. As the term transmission electron microscope implies, the image is formed by electrons, which pass through the specimen and strike either a fluorescent viewing screen or photographic material. The bright areas of the image correspond to less dense or open areas of the specimen through which the electrons pass relatively unimpeded. Similarly, dark areas correspond to areas of the specimen that impede the transmission of electrons.

One might first reason that the electrons that strike solid areas of the specimen are absorbed and are, therefore, not able to contribute to the image. This is only true for a small percentage of incident electrons; i.e., for an electron to be absorbed it must transfer all its energy, and thus velocity, to the impinging material. This is accomplished by means of a series of inelastic collisions with electrons of the specimen. The results of an inelastic collision include reduced velocity of the primary electron as well as a change in its direction.

In order for a primary electron to be fully absorbed, there must be sufficient inelastic collisions to reduce its velocity to zero. In practical terms, this can only be accomplished if the specimen is much thicker or denser than normal biological sections. In fact, normally the copper grid bars are the only areas to fully absorb primary electrons. Primary electrons involved in inelastic collisions with the specimen may not be fully stopped; however, they do change direction and are said to be *scattered*. These scattered electrons are, for the most part, absorbed by parts of the microscope column and do not contribute to the final image.

Not all scattered electrons are absent from the final image. If the angle of scatter is not large enough, they will not be absorbed by the surrounding column but instead will contribute to the final image. The problem with allowing these electrons to contribute to the final image is that they follow completely random paths down the column and thus tend to stray into areas of the image that should appear dark. This results in a "washing out" or loss of contrast in the image.

The *objective aperture* is used to absorb these scattered electrons and thus prevent them from contributing to the final image, resulting in a higher contrast image (Fig. 6.13). The objective aperture is usually constructed from a disk of platinum or a thin film of gold with an aperture size ranging from 5 to 100 μm, depending on the amount of contrast desired and objective lens design. The transmitted and forward scattered electrons are combined to form a diffraction pattern at the back focal plane of the objective lens where the objective aperture removes some of the scattered electrons. A magnified image (approximately 30 ×) is then formed at the image plane. By use of a series of imaging lenses – usually a diffraction, intermediate, and two projector lenses – either the image or the diffraction pattern can be projected onto the viewing screen or film. (The diffraction pattern is used mostly by material scientists, because it contains crystal structure information about the specimen.)

Earlier, it was stated that the practical resolution of a TEM is somewhat less than the theoretical limit of resolution due to certain physical limitations. One such limitation, which is dependent on lens design, is the spherical aberration coefficient (C_s) of the objective lens. (All lenses suffer from this aberration, but it is the value of C_s for the objective lens that affects resolution the most.) This is a measure of a lens's ability to focus all electrons passing through the lens (parallel to the principal axis) to the same spot, as would be the case in a perfect lens [Fig. 6.14 (top)]. In practice, it is found that electrons passing through the lens farthest from the principal axis are subject to greater curvature and will pass through the axis at different points, causing a blurred image [Fig. 6.14 (bottom)].

Spherical aberration can be greatly reduced by eliminating the peripheral electrons. One method of doing this is to use an objective aperture to stop down the lens physically, i.e., block the peripheral electrons from passing through the lens. Unfortunately, this also reduces the information-gathering ability of the lens, thereby limiting the resolving power of the microscope. Usually an optimal size of objective aperture is recommended by the manufacturer, which represents a good compromise between contrast and resolution.

Another method of reducing peripheral electrons through the objective lens is to limit the angular aperture of the objective lens (α_0) (Fig. 6.15) (defined as the area of the lens through which all electrons must pass). One way of doing this is to reduce the angle of illumination (α_1) by having the electrons travel as parallel as possible to the principal axis. This can be done by simply using the second condenser lens in an over or underfocused condition. Earlier it was stated that if the C2 lens is used in a focused condition (i.e., the light is focused to a small bright spot), the image appears blurred and lacking in detail. This can be explained by examining three different conditions of the C2 lens: focused, underfocused and overfocused.

Figure 6.16 (left) shows the situation where C2 is focused, i.e., its focal length is such that the object plane for the C2 lens corresponds with the image plane for the C1 lens. This condition results in the smallest, most intense spot at the specimen as well as the largest angular aperture of illumination (α_1) and therefore, the largest angular aperture for the objective lens also. Since α_0 is at a maximum, this condition will allow the largest number of peripheral electrons to pass through the objective lens, thereby limiting resolving power. This is also the only condition in which the outline of the C2 aperture can be seen and an image of the filament will be in focus. This condition is used to center the adjustable C2 aperture and to center and saturate the filament accurately.

Figure 6.16 (center) shows the situation where the C2 lens is overfocused, i.e., its focal length has been shortened. It can be seen now that α_1 has decreased and that the condenser aperture is no longer the limiting factor. Because α_1 has decreased, α_0 will

Figure 6.14. **Top.** Perfect lens. **Bottom.** Lens with spherical aberration.
Figure 6.15. Relationship between numerical aperture of objective lens and spherical aberration. A lens with α_{02} allows fewer peripheral electrons to pass than does one with α_{01}; therefore spherical aberration is less with a smaller α_0.
Figure 6.16. Relationship between focal length of condenser lens and angle of illumination (α_1). **Left.** C_2 focused. **Center.** C_2 overfocused. **Right.** C_2 underfocused.

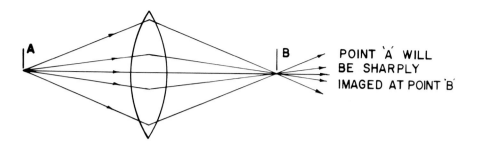

POINT 'A' WILL
BE SHARPLY
IMAGED AT POINT 'B'

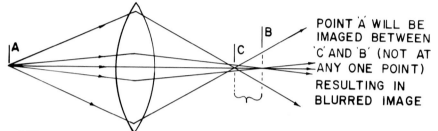

POINT 'A' WILL BE
IMAGED BETWEEN
'C' AND 'B' (NOT AT
ANY ONE POINT)
RESULTING IN
BLURRED IMAGE

Figure 6.14.

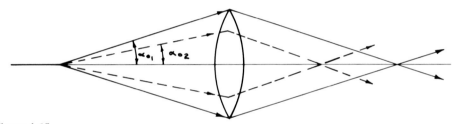

Figure 6.15.

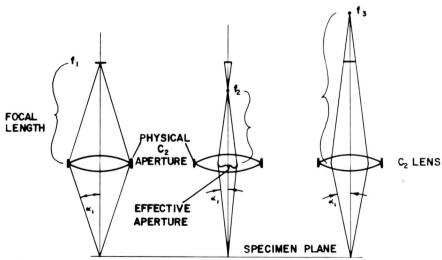

Figure 6.16.

also be smaller, thereby reducing the number of peripheral electrons through the objective lens.

Figure 6.16 (right) shows the C2 lens in an underfocused condition, i.e., where its focal length has been extended beyond the image plane of the C1 lens. As in the overfocused condition, the angle of illumination has been reduced, thereby limiting peripheral electrons through the objective lens.

Another benefit of working with an overfocused C2 lens is that the beam coherence increases as the angle of illumination decreases. Briefly, this means that the electrons from the gun will all hit the specimen at close to the same angle, i.e., rays are nearly perpendicular to the specimen surface when α_1 is small. This condition will give rise to a well-defined interference pattern, and it is one such pattern, the Fresnel diffraction pattern, which is very important for image contrast and as a focusing aid.

Working with the C2 lens in an out-of-focus condition will result in reduced spherical aberration and better beam coherence, both of which will give a better image. However, the intensity of illumination varies inversely as the square of the angle of illumination. This means that brightness falls off dramatically on either side of C2 focus, which means that emission current must be increased to achieve sufficient illumination for viewing or photography. The result is a much improved image but reduced filament life. How much this compromise will affect filament life depends on how well the operator adapts to low light conditions; i.e., the room should be totally dark, and the operator should use minimum brightness required for comfortable viewing.

IMAGE VIEWING AND RECORDING

A tremendous amount of structural and chemical information about a specimen can be gathered by looking at the various by-products resulting from the collision of the primary beam with the specimen, namely, backscattered electrons, secondary electrons, X-rays, and transmitted electrons. In a clinical application it is transmitted electrons, both elastically and inelastically scattered, that are important, since they combine to form the final image. If this image (which is the most important component in the TEM) is poor, all the physics, chemistry, and machining that went into developing the microscope were wasted.

The most common method of viewing an image is via direct imaging on a phosphor screen which can be viewed through a leaded glass window or magnifying binoculars. The phosphor on the screen, when struck by an imaging electron, will emit radiation with a wavelength in the visible light spectrum, usually green light.

Another way to view the image directly is by using a television (TV) camera instead of the screen and viewing the image directly on a TV monitor. This has the advantage of extra magnification through the TV camera plus the convenience of instant recording and play-back on a video recorder. The disadvantage is the loss of image quality, especially at high magnification, due mostly to inherent electronic noise found in all TV cameras. This is becoming less of a problem with the development of better high-resolution cameras and digital image storage.

At present the most common method of image recording is on photographic film, which can come in different formats, i.e., 35 mm, 70 mm or $3\frac{1}{4}$ in. × 4 in. (85 mm ×

102 mm) sheets. The film of choice depends on both the model of TEM being used (some manufacturers stress one format over another) and the operator's darkroom facilities.

IMAGE OPTIMIZATION

The ability of any TEM to achieve its theoretical limit of resolution depends greatly on the precision used by the manufacturer when making the lens pieces, as well as when designing and manufacturing the electronic parts. A well-built TEM will enable the microscopist to achieve the machine's theoretical resolution limit for 20 years. Aside from selecting a well-engineered machine, the best performance is only possible if attention is given to specimen preparation (Chapters 1 to 3) and proper mechanical and electrical alignment.

Aligning a TEM column involves positioning of the lenses so that they are centered around one vertical axis. If this axial alignment is achieved, it should be possible to change the strength of one or a combination of lenses (as in focusing or a magnification change), without displacing the image from the center of the viewing screen. Some rotation of the image about the center of the screen is normal, due to the inherent characteristics of magnetic lenses.

A quick check of the alignment can be done by focusing an image at low magnification while keeping the illumination (or spot) centered with the deflection controls. It should then be possible to increase the magnification to maximum with the image remaining centered on the viewing screen. The ability of the image to remain centered during a magnification change is a measure of the accuracy of the imaging lens's alignment, namely that of the diffraction, intermediate, and projector lenses.

The other critical test of the alignment is done by focusing the image. If the objective lens is not centered, the image will sweep in an arc about some point other than the center of the screen. This sweeping effect is worse at higher magnifications and can be very annoying when viewing the image through binoculars, since it can move right out of view. If properly aligned, the image at the center of the screen will rotate slightly about the center, but should not be displaced.

Depending on the age of the microscope, the role of the operator in the alignment procedure can vary immensely. Until the mid-1970s, the lenses had to be physically aligned by moving them with a series of alignment screws. On these early machines, the operator or service engineer was responsible for the alignment of each lens and, depending upon experience and the degree of precision required, the entire alignment could take hours. Normally this was only necessary if the column was dismantled for cleaning and then reassembled.

In the mid-1970s, manufacturers started to build microscopes with permanently aligned lenses; i.e., the lenses are centered as part of the manufacturing process and then fixed in this position. These types of factory-aligned columns are convenient for the operator, since no major alignment is required, but the alignment is never as accurate as a manual alignment. This is due primarily to limitations encountered during assembly and slight misalignments during shipping. For this reason, manufacturers have added a series of alignment-correction coils at various locations in the column.

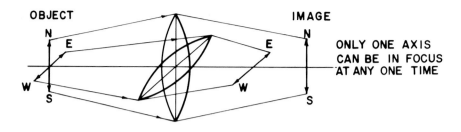

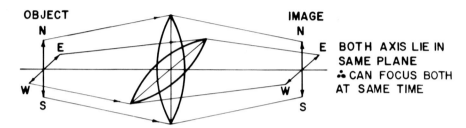

Figure 6.17. **Top.** Asymmetrical lens (astigmatic). **Bottom.** Perfectly symmetrical lens (no astigmatism).

These coils act as small magnetic lenses and allow the operator to correct any small errors in the factory alignment. Use of these coils results in a near-perfect alignment, as with a manual type alignment, but now it requires only minutes and very little skill.

Today's microscopes all come equipped with a built-in computer, which is used to control the microscope's various systems. All values of lens and correction coil currents can be stored on a floppy disk, which can be used to realign the microscope instantly, should this be necessary. This computer advancement is a mixed blessing in that anyone, regardless of level of expertise, can realign the microscope even if he or she in unfamiliar with electron-optics theory. This reduces both the skill required to operate the instrument and the necessary level of understanding of optics, which at the same time can result in a lowering of the operator's ability to recognize and correct simple faults.

Along with the beam-alignment coils, the TEM usually contains other correction coils called stigmators. These operator-controlled coils are used to correct what is termed astigmatism, which is an inherent property of any lens and, put simply, is the ability of the lens to magnify equally in all planes. If the lens is not perfectly symmetrical, which is normally due to manufacturing limitations, then the N–S and E–W axes will not be in the same plane and will cause the formation of a blurred image [Fig. 6.17 (top)]. If a lens is perfectly symmetrical, then the N–S and E–W axes of an image will lie in the same plane [Fig. 6.17 (bottom)].

Stigmators, or correction coils, are usually provided to correct the astigmatism of condenser, objective, and diffraction lenses. Condenser and diffraction lens stigmators are provided to enable the operator to form a circular spot instead of an ellipse, which is the case in an astigmatic lens. This is important, especially in high-resolution work, where a circular spot of even illumination and maximum brightness is essential (Fig. 6.18).

Figure 6.18.
    A. Filament image with condenser lens astigmatism corrected.
    B. Filament image without condenser lens astigmatism corrected.

Correction of objective lens astigmatism is the most critical, since this lens forms the first image that must also be in the correct plane for the magnifying lenses. If any astigmatism exists, it will not be possible to focus the N–S and E–W planes together, and the result will be an image with perpendicular planes at different degrees of focus. This will be seen as a blurred image lacking detail. The easiest way to correct this type of astigmatism is to make use of the Fresnel fringe (Fig. 6.19), an interference phenomenon, which appears bright when the image is underfocused and dark when overfocused. If the image is astigmatic, half the Fresnel fringe around a hole will be dark and half will be bright (Fig. 6.19A). The objective stigmators should be adjusted until the entire fringe is either bright or dark (Fig. 6.19B).

An experienced operator can correct objective lens astigmatism by observing any fine-grain structure at high magnification. In this case the stigmators are adjusted until the image is very sharp and equally focused in all directions (Fig. 6.20).

Once the instrument has been properly aligned and the stigmators adjusted, there are still some steps to be taken in order to optimize the image. Since these all involve selecting proper operating parameters, the operator's experience and basic knowledge of optics will determine the degree of success. For example, the correct second condenser and objective apertures must be selected and centered, the size of which will depend on the type of microscope (analytical or clinical) and on the resolution desired.

Another operator-selected parameter that affects image quality is the accelerating voltage, or high tension (HT). It was stated earlier that resolving power depends on the wavelength of light or energy used, and it is for this reason that an electron beam, with a very short wavelength, is used in the TEM. As the accelerating voltage is increased, the velocity of the electrons flowing down the column is increased and using de Broglie's relationship between the velocity and wavelength of an electron, it is seen that the wavelength of an electron decreases as the HT increases. Therefore, for high resolution work the instrument should be used at the highest value of HT available. Some of today's commercial high-resolution TEMs are capable of HT values up to 400 kilovolts, while some custom-made TEMs have HT values greater than 1 megavolt!

The problem with using high HT values in a clinical setting is the lack of contrast that results. As their energy (or velocity) increases, less electrons are absorbed by the specimen but are instead scattered, thus washing out the image. For most biological specimens, a value of 60 kV gives a good compromise between resolution and contrast.

The aperture sizes, the value of HT, and the C1 lens setting (spot size) are parameters usually selected once for the type of work being done and then left alone. In fact, most operators will never have to worry about these adjustments. However, every operator is responsible for the correct operation of the C2 lens (illumination intensity), and it is this parameter that is most often set incorrectly.

Most operators, including even experienced ones, tend to focus the C2 lens to a small intense spot, since this gives a nice bright image that can be seen easily without having to work in a dark room. By doing so, the angle of illumination is increased and image resolution reduced. In the earlier section on condenser lenses, it was shown that resolution increases greatly by reducing the angle of illumination at the specimen and this is accomplished by using C2 lens in an under- or overfocused condition. The drawback here is the necessity of working in a darkened room.

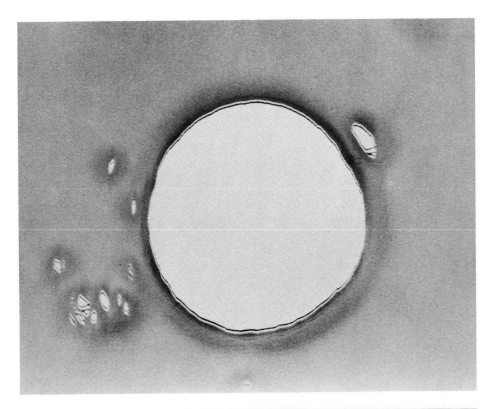

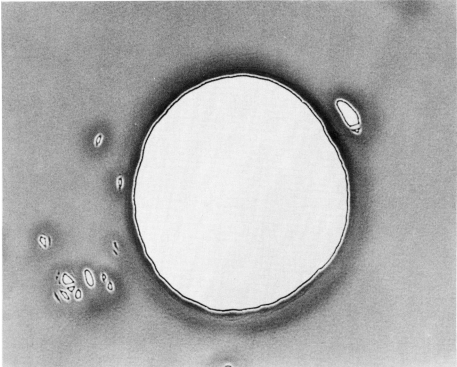

Figure 6.19. *Objective lens astigmatism.*
  A. Uncorrected.
  B. Corrected.

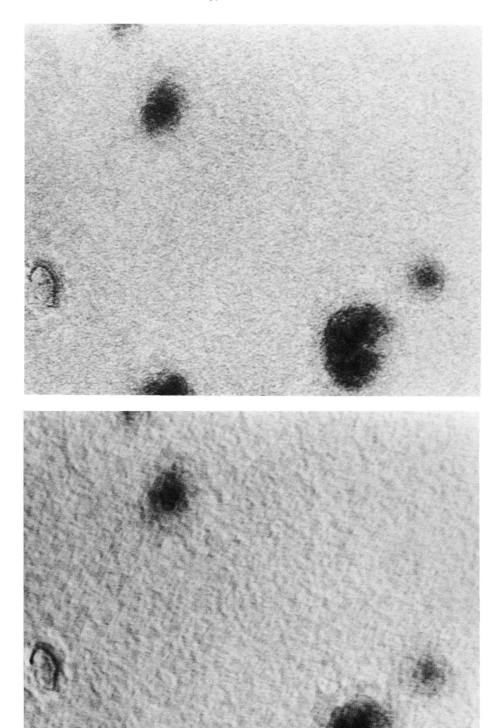

Figure 6.20. *Correction of objective lens astigmatism using grain structure.*
  A. Corrected astigmatism.
  B. Uncorrected astigmatism

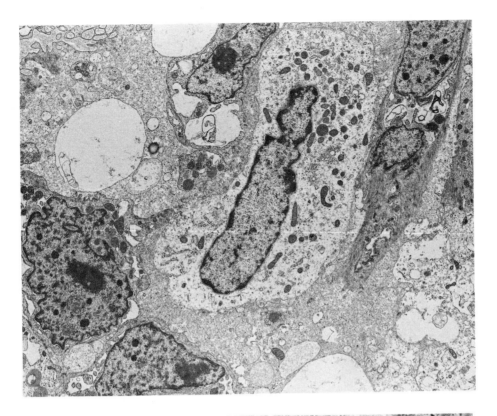

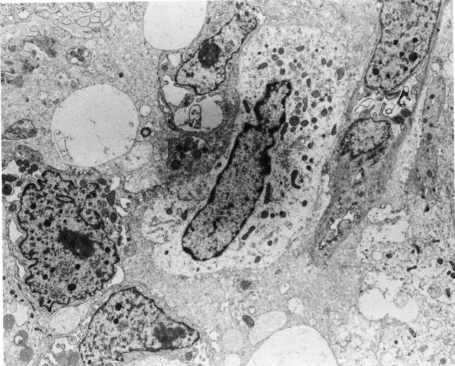

Figure 6.21. *Effect of oil and water vapors on specimen.*
  A. Uncontaminated specimen.
  B. Same specimen following prolonged exposure to electron beam .

The other problem with a small and intense beam is specimen contamination. In a vacuum system, regardless of how well it is designed, there are residual oil and water vapors that arise from pumping oils and vacuum-seal grease. These vapors, which are mostly hydrocarbons, when exposed to an electron beam, decompose to carbon and hydrogen gas. The gas is readily pumped away, but the carbon remains as a coating on the specimen (Fig. 6.21). This coating will actually build up on edges and projections, thus changing their shapes and causing increasingly poor resolution.

Contamination can be reduced by simply working with C2 in an overfocused condition, which will reduce the intensity of the beam and therefore reduce the effects caused by the interaction of the beam with vapors and the specimen. Another obvious method of minimizing this problem is to reduce vapors in the column, or at least near the specimen. This has been accomplished to a large extent by designing so-called "clean columns" with very few rubber "O" rings (and thus less vacuum grease) and by the use of efficient pumping systems with vapor traps. These measures, along with the use of ion-getter pumps to maintain the specimen area at ultra-high vacuum levels, have greatly reduced contamination rates. By partially isolating the photographic material from this "clean vacuum" most of the water vapor, and thus etching, has been eliminated.

Even with the above design considerations there will always be some residual gases to cause contamination and etching, and anyone doing high-resolution work or work on very beam-sensitive material, such as latex or biological samples, must take additional precautions to further reduce the detrimental effects caused by contamination and etching.

One of these devices is a cold trap, or anticontamination device, which is simply a cooled surface placed as close to the sample as possible. Since vapors will condense more readily on a cool surface than on the specimen, this surface (usually metal plates or blades surrounding the specimen) is kept at the liquid nitrogen temperature of $-150°C$. Use of a cold trap plus proper use of the condenser lens, i.e., working as far overfocused as is practical, can reduce the effects of contamination and etching to undetectable levels.

MAINTENANCE

All microscopes require routine cleaning, e.g., especially of the Wehnelt cylinder after filament change, and the aperture and specimen holders, in addition to that provided by a service contract or a preventive maintenance schedule. The following precautions are important for top TEM performance:

- Have all cleaning materials at hand before opening column.
- Work in a clean, ventilated room.
- Use a clean work bench and good illumination.
- Wear clean, lint-free gloves when handling "vacuum" parts.
- Use clean solvent.
- Do not mix materials of different compositions in cleaning baths.
- Before mounting parts, inspect them under a dissecting microscope.
- Wrap cleaned parts in aluminium foil for storage.

Table 6.1. *Problems encountered in use of electron microscope and their possible solutions.*

Problem	Possible causes	Action to take
Microscope cannot be switched on	No mains voltage	Check fuses
	Standby switch off	Check switch
	Mains protection circuit in operation due to electronic fault	Call for service
No rotary pump activity	Blown fuse	Check fuses
Oil-diffusion pump off	Insufficient cooling water	Check water source Clean filters
	Heating element cold	Check fuse Replace heater
Excessive time taken for rotary pump to pump out column or camera	Vacuum leak	Check all hoses, couplings, and removable ports on column (e.g., camera, electron gun)
	Wet film in vacuum	Remove film and desiccate it outside the microscope
	Rotary pump oil saturated with water	Open ballast port on pump for 1 hour Change oil and aluminium oxide pellets
High vacuum poor	Small vacuum leak	Check "O" rings on camera port, etc. Call for service
	Wet film	Desiccate film better before putting in microscope
	Humid ambient air	Use dry N_2 gas when venting column, e.g., during film or filament change
	IGP old	If vacuum does not improve overnight replace pump
No illumination although HT is on	Filament blown	Check emission current. If none, change filament
	Something blocking beam	Remove apertures and specimen Check camera and shutter positions Call for service
	Filament not aligned	Align filament and gun
	C2 lens left in overfocus position	Focus C2 lens
Illumination present, but not very bright	Gun misaligned	Align filament and gun
	Emission current too low	Increase emission
	Filament undersaturated	Image the filament (focus C2), then saturate
	C2 or objective apertures too small or misaligned	Check aperture size and recenter
	Spot size (C1 lens) to small	Check operator's guide for optimum setting of C1

(Table 6.1 cont'd on p. 132)

Table 6.1 (cont'd)

Problem	Possible causes	Action to take
Loss of illumination as filament is saturated	Filament not centered	Remove filament assembly and recenter
Illumination sweeps across the screen when focusing C2	C2 aperture not centered	Center C2 aperture
Specimen drifts	Grid not clamped to holder properly	Remove grid, check visually with loupe
	Section has come loose from grid	Try another specimen
	Dirty grid or holder	Try another grid and holder Clean holder
	Specimen stage or objective aperture rod dirty	Call for service
Specimen damages easily	Specimen too wet	Dry specimen in vacuum or under heat lamp
	Too much water vapor in column	Check liquid N_2 level in cold trap Wet film in column; remove it
	Vacuum leak at specimen level	Check "O" rings on objective aperture and specimen rods Call for service
	Specimen very electron-sensitive	Work with lower level of illumination (overfocus C2 more)
Image not sharp	C2 lens at focus	Work overfocused
	High objective lens astigmatism	Correct with objective lens stigmators
	Dirty specimen	Try another specimen
	Dirty or misaligned objective aperture	Realign or clean holder and replace aperture
Image sweeps across the screen when focusing	Objective lens not centered	Align objective lens
Centered image moves off center when changing magnifications	Image lenses not aligned	Align image lenses

• Whenever possible, use household abrasive (e.g., Vim or Jif), ethanol, and a neutral cleaning fluid and soap (e.g., Extran) instead of Wehnol paste and fluorocarbons.

To Clean Heavily Contaminated Metal Parts:

1. Rub with Vim; rinse in tap water.
2. Ultrasonicate 5 minutes in 5% Extran, rinse 5 minutes in distilled water.
3. Ultrasonicate 5 minutes in ethanol.
4. Ultrasonicate in three changes of 99.8% pure ethanol, 2 minutes in each.

5. Air-dry thoroughly at 80°C with infrared lamps or in a vacuum oven.
6. Jet dust immediately before installing in EM.

To Clean Wehnelt and Anode:

1. Ultrasonicate for 30 minutes in 6% sodium or potassium hydroxide in distilled water (prevent skin contact).
2. Wash in tap water.
3. Ultrasonicate in Extran, etc. and dry as described previously.

To Clean Less Contaminated Metal Parts:

1. Ultrasonicate 5 minutes in 5% Extran; rinse in distilled water.
2. Ultrasonicate in ethanol and dry as described previously.

To Clean "O" Rings and Other Synthetic Parts:

1. Ultrasonicate 5 minutes in ethanol.
2. Rinse in fresh ethanol and air-dry.

Trouble-Shooting Guide

To avoid unnecessary down time as well as potentially costly service visits, the user must be able to recognize and correct simple routine problems. Some of these may be the result of improper use, while others may be true mechanical or electrical faults. Table 6.1 lists a numbers of problems that a TEM operator may encounter during normal operation. It also gives suggestions for solving these problems (before calling the service person!).

7

• *Photography*

INTRODUCTION

After material has been prepared for viewing with the electron microscope, the most important step is to make a permanent photographic record of the findings. There are two main reasons for this: (1) the graininess of the phosphorescent viewing screen limits resolution, and (2) observation time is short due to progressive specimen contamination by the electron beam. Other reasons are (1) details not observed by the microscopist are often seen on a print, (2) two areas of a specimen cannot be compared on the EM screen, (3) it is very difficult to relocate a specific area on a grid, (4) photographs are essential for publication and/or to substantiate findings (sometimes a second opinion is required), and (5) a microscopist might want to make comparisons with previously published findings. A common mistake of beginners is to spend too long viewing the specimen. It is far more productive to take many pictures that can be examined carefully later; photographic film captures more detail than the microscope screen and is relatively inexpensive.

Proper control of exposure conditions is very important if one is to produce good-quality electron micrographs. One must follow the operator's manual for basic operations of camera, shutter mechanism, and metering system. There are, however, some general guidelines to follow in order to avoid common problems such as uneven illumination, blurred images due to specimen drift or astigmatism, damage to the specimen from the electron beam, and low-contrast negatives.

The microscope must be properly aligned, astigmatism corrected, and the grid holder and objective aperture free of contamination. Improper alignment is a frequent cause of unevenly exposed negatives. A contaminated grid holder will cause the specimen to both drift and to go in and out of focus. A contaminated objective aperture will cause intermittent astigmatism, astigmatism that cannot be corrected, and/or an intermittent out-of-focus condition.

The type of specimen being examined could determine selection of both accelerating voltage and objective aperture size. Selecting a low accelerating voltage (e.g., 40 kV) or small objective aperture (e.g., 30 μm) for low-contrast specimens will result in higher contrast in the negative. Conversely, for high-contrast specimens higher accelerating voltage (e.g., 80 or 100 kV) or a larger objective aperture (e.g., 60 μm) will reduce contrast. Higher accelerating voltage will also result in better resolution, less beam damage to the specimen, and significant reduction in specimen drift.

135

The beam in an electron microscope is a very intense source of radiation, which can damage the specimen by a complex mixture of heating and ionizing effects. Sections may sag or collapse when they are first exposed, and the irregular dissipation of heat causes specimen drift. Most EM films require a 1–3 second exposure so any movement or drift of the specimen during this time will cause a blurred image on the negative. Sections will move less when they are supported on the grid by a plastic film such as Formvar or Butvar, but this reduces contrast and resolution and can be a source of contamination. Sections collected on smaller mesh grids (400 mesh as opposed to 200 mesh) show less movement, but areas of the specimen immediately beside a grid bar often show very acute drift. Contaminated, holey, or damaged specimens will also drift and be impossible to photograph well.

Selecting the desired area to photograph at low magnification will minimize local beam damage. Low-magnification scanning stabilizes the section and reduces the amount of movement when areas are later magnified. Some workers describe low-magnification scanning of thin sections as "ironing the section."

Once an area on the specimen has been chosen to photograph, one must focus the image and adjust illumination. It is important to first focus the oculars of the binocular viewing microscope using a mark on the focusing screen, pointer, or diffraction beam stop. Focusing at low magnification is more difficult than at high magnification and for this reason, some electron microscopes are equipped with a wobbler. This causes a deflection of the beam such that, when the objective lens is out of focus, the image appears to be double. Correct focus eliminates the double image. A slightly under-focused image produces an increase in contrast which is pleasing to the eye, but care should be taken not to underfocus to the point where there is loss of sharpness.

More intense illumination is used for viewing specimens through the binoculars than is required for photographic exposures, but the final focus should be done at the lowest possible illumination. After focusing the image one must reduce illumination, using the condenser lens control, to an exposure meter value (usually indicated by exposure time) which has been predetermined by a test film. It is important that the Wehnelt bias current (filament emission) has been set to a value high enough to allow exposures to be made with the condenser lens in the overfocused position. Any change in either bias current or magnification after the image has been focused will necessitate refocusing. After reducing illumination, check for specimen drift and allow it to stop before activating the shutter. Vibration from either the rotary vacuum pump or from someone resting against or bumping the microscope must be avoided since this can cause blurring in the recorded image.

Photographic films are made by coating glass, plastic (polyester or acetate), or paper with an emulsion of fine halide crystals dispersed in gelatin. The currently accepted mechanism of image formation on silver halide films was conceived by Ronald W. Guerney and Nevill F. Mott (1938), and although they used the photon as an energy source, their findings also apply using a beam of electrons. When an energy source is used to excite a silver halide crystal, electrons are liberated and move freely to sensitivity sites within the crystal where they are trapped. The negative charge attracts positively charged silver ions to form metallic silver atoms. When three or four silver atoms have clustered in the same location, a stable development center (latent image center) forms, which acts as a starter point for the reducing agents in the developer to convert the remaining silver halide of the crystal to silver.

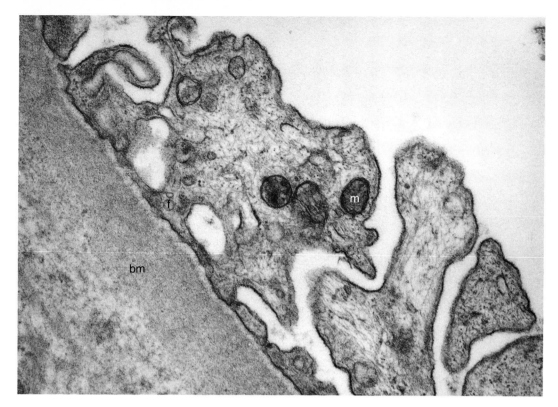

Figure 7.1. *Electron micrographs showing effects of accelerating voltage on negative contrast.*
A. Section of human renal capillary taken at 50 kV. Podocyte foot processes (f); external lamina (bm); mitochondria (m).

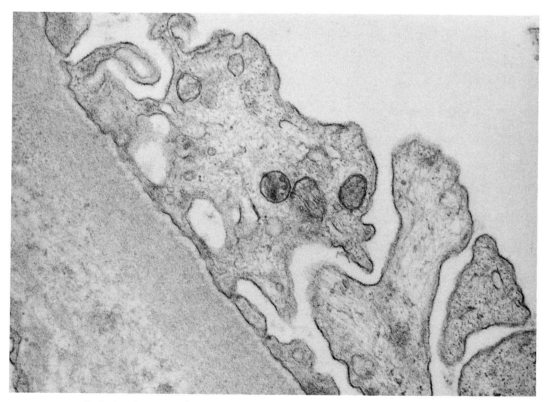

B. Reduced contrast shown in same section micrographed at 80 kV and printed on the same grade of paper.

A major difference between photons and electrons is that electrons have much more energy. While it might require 10 photons to excite one silver halide grain, one electron can excite 10 silver halide grains. For this reason, only fine-grain films are suitable for electron microscopy. Also, because of the increased energy of electrons, it is necessary to have a relatively thick emulsion (5–10 μm) in order to slow the electrons, thus increasing resolving power. In general, EM films have high contrast, short exposure latitude, and slow speed. Most films are capable of recording every incident electron (perfect recorders) and any "noise" (graininess) that occurs is due to random fluctuations of electrons and not to the grain of the emulsion. This gives high resolution in the print even after the negative is enlarged by a factor of 8 to 10. Films used for EM have a high signal-to-noise ratio [expressed quantitatively by Kodak scientists as the detective quantum efficiency (DQE)]. Films with a DQE of 1 are perfect recorders (Kodak Pamphlet No. N-921).

Films exposed to electrons continue to increase in contrast with exposure. A denser negative has more contrast, and the signal-to-noise ratio is improved. This is because image contrast increases linearly with exposure (more electrons), while noise increases less rapidly by the square root of exposure. Contrast refers to the difference between the most dense and least dense areas of a negative and can be increased by reducing magnification at the electron microscope (more electrons). Increasing development time or temperature also increases contrast, but this increases noise. Decreasing accelerating voltage on the electron microscope will increase contrast (Fig. 7.1A, B) but will decrease resolution. To check density, place a negative, emulsion side down, on a clean newspaper; in good light, the newspaper print should just be visible through the negative.

Silver halide emulsions are normally only sensitive to wavelengths at and below blue light (450 nm) unless dyes are added to change the color sensitivity. Most EM films are blue-sensitive, which permits handling under safelights with dark brown or green filters. Safelight recommendations are given on each box of film and must be followed.

Film formats used in EM include 35-mm rolls and $3\frac{1}{4}$ in. × 4 in. (83 mm × 102 mm) sheets. Both give adequate image resolution but 35-mm roll films are less expensive and easier to handle than sheet film. Also their small size results in lower costs for developer and fixer. The advantages of sheet film are that the larger format requires less photographic enlargement and, because most microscopes can record individual identification numbers on sheet film, negative storage and retrieval are simplified. In order to protect the EM column, films should be degassed in a separate vacuum chamber before being placed in the microscope. Most EM films have a support made of polyester (e.g., Estar, Mylar, Cronar) which does not emit significant amounts of volatile gases; the main source of vapor comes from atmospheric water absorbed by either the silver halide emulsion or protective coating of gelatin. Emulsions on glass plates are available but are rarely used because they are expensive, present additional handling and storage problems and, in the author's opinion, produce results indistinguishable from those using acetate film.

FILMS FOR ELECTRON MICROSCOPY

- Kodak Electron Microscope Film SO-281, polyester (35-mm roll).
- Kodak Technical Pan Film 2415, polyester (35-mm roll), panchromatic.

- Kodak Electron Microscope Film 4489, polyester [$3\frac{1}{4}$ in. × 4 in. (83 mm × 102 mm) sheet].
- Kodak Electron Microscope Film SO-163, polyester [$3\frac{1}{4}$ in. × 4 in. (83 mm × 102 mm) sheet].
- Ilford EM film, polyester [$3\frac{1}{4}$ in. × 4 in. (83 mm × 102 mm) sheet].
- Agfa Scientia 23D56, polyester [$3\frac{1}{4}$ in. × 4 in. (83 mm × 102 mm) sheet].

FILM PROCESSING – GENERAL COMMENTS

Processing films involves three steps: developing, fixing, and washing. The purpose of development is to reduce silver halide to metallic silver. As mentioned earlier, contrast can be increased by increasing development time, temperature, or frequency of agitation. Also, different developers or different concentrations of the same developer can produce entirely different densities in negatives, given the same electron exposure. Bear in mind, however, that increasing density in this manner can also increase noise, so a better alternative is to either increase microscope exposure time or decrease microscope magnification (both result in more electrons hitting the emulsion). In the latter case, subsequent enlarger magnification will not reduce contrast. Film manufacturers' recommendations may need to be modified for different microscopes.

Since electrons have such high energy, latent image density continues to increase after exposure; therefore, films should be developed as soon after exposure as is practical. Figure 7.2 shows prints from negatives exposed 24 hours apart. Incomplete film drying enhances this effect (as was the case in these micrographs).

Developing agents are organic compounds (phenols or amines) and include: metol (4-methyl-aminophenol sulfate), hydroquinone, phenidone (1-phenyl-3-pyrazolidone), and amidol. Sodium or ammonium sulfite is added to extend the life of the developer by forming soluble compounds with oxidation products. Alkalis, such as sodium hydroxide, are added to increase pH to above 8.5, thereby increasing development speed. Some developers contain soluble bromides (called "anti-foggant") to reduce development of unexposed silver halide.

Commercial developers come in two forms, powders and liquid concentrates. Powder developers have an unlimited shelf life if stored properly in their unopened packets but, once dissolved in water, they should be stored in tightly stoppered, dark brown bottles. Powder developers are not homogeneous and for this reason, the packets cannot be subdivided to make smaller volumes of working developer. Liquid concentrate developers have a short shelf life compared with powders but, because any desired volume of working developer can be made quickly, they are very convenient and time saving. Oxidation of working developer will occur even in air-tight bottles and, in most cases, developer should be used within 6 months of preparation. Developer can be reused if kept in a covered container. A regular schedule for replacement should be established (e.g., approximately 10 days or 500 plates). Roll-film developer should be discarded after one use. Money spent replacing developer is minor compared with the cost of retaking unusable films. Following development, film is rinsed in water (do not use stop bath as it can cause "mottling" on EM film). This neutralizes and removes developer and extends fixer life.

Fixing stops development of the exposed silver grains and removes unexposed

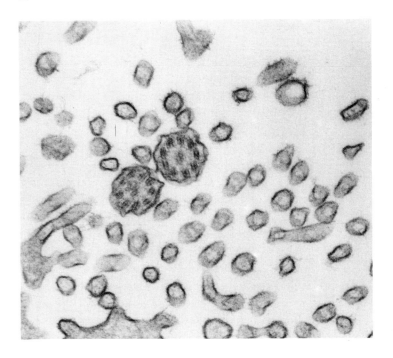

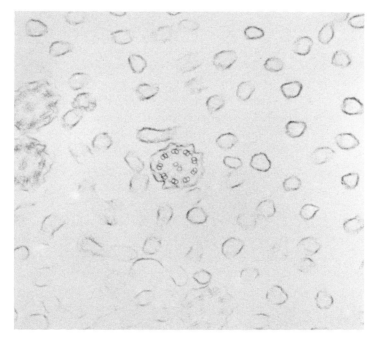

Figure 7.2. *Electron micrographs of cilia showing effect of electrons on photographic emulsion.* Top micrograph was taken 24 hours after bottom micrograph. Both, on the same film, were developed at the same time shortly after top micrograph was exposed. Both were printed on same grade of paper and the negatives exposed in enlarger for same length of time.

silver halide. Fixers contain 20–30% sodium or ammonium thiosulfate in a mildly acidic buffer. They also contain sulfuric acid, which hardens the emulsion and protects it from handling abuse. Fixing time (usually 2–3 minutes) should be twice the time required to clear the unexposed silver. Fixer can be reused but should be replaced each week to 10 days, depending upon the number of films that have been developed. It is wise to use two fixing baths – 2 minutes in the first, followed by 1–2 minutes in the second. After 500 plates or 10 days, the first fixer is discarded, the second fixer becomes the first, and a fresh one is prepared as fixer 2. One can inspect EM film under a safelight; it will change from white to clear when about half the time required to fully fix has elapsed.

Finally, films must be washed and dried. They are washed for 30 minutes, in running tap water at 20°C to remove fixer and by-products of development. Residual halide–fixer complexes can stain the film and also cause image instability. After washing, the film is rinsed with a wetting agent (e.g., Kodak Photo-Flo) to promote even drying and eliminate water spots. The film is then dried in a clean cabinet under warm, circulating air.

In photography, cleanliness is essential if high quality is important. Counter tops must be cleaned regularly with a damp sponge to reduce dust. Sinks, tanks, beakers, and glassware must be washed to remove photochemical residues. Plastic reels should be taken apart and rinsed after the Photo-Flo rinse step and dried thoroughly before reuse. Dilute Photo-Flo should be changed (and the container washed) once a week to prevent mold growth. To reduce dust, hands and darkroom equipment should be dried using cotton rather than paper towels.

Processing EM Sheet Film $3\frac{1}{4}$ in. $\times$ 4 in. (85 mm $\times$ 102 mm)

Chemicals and Equipment

- Developer appropriate for film used.
- Fixer.
- Kodak Photo-Flo.
- Graduated plastic cylinders or beakers – 4,000 mL, 1,000 mL, 100 mL.
- Kodak developing tanks (6), black rubber – 3.8 L, 5 in. $\times$ 8 in. $\times$ $7\frac{3}{4}$ in. (127 mm $\times$ 200 mm $\times$ 194 mm).
- Sheet film developing racks (2), clear plexiglass, 20-film capacity.
- Darkroom timer with luminescent dial.
- Safelight suitable for blue-sensitive films.
- Storage envelopes of transparent plastic for $3\frac{1}{4}$ in. $\times$ 4 in. (83 mm $\times$ 102 mm) sheet films.

Method

1. Prepare developer and fixer according to manufacturers' instructions and fill tanks. Do the following steps under appropriate safelight illumination.
2. Remove exposed sheet films from electron microscope and load into plexiglass rack.
3. Immerse loaded rack into developer tank; set timer. Raise and lower rack several times (without lifting film out of developer) to remove air bubbles at film surface. Agitate 5 seconds every 30 seconds by raising and lowering the rack.
4. Transfer developed films to a second tank containing running water at 20°C and rinse approximately 1 minute.
5. Transfer to a third tank containing fixer. Agitate constantly for 2 minutes.
6. Place films in a second fixing bath for 1 minute, then into a tank of running water at 20°C. Rinse for 20–30 minutes.
7. Immerse in a sixth tank containing Photo-Flo and agitate about 30 seconds.

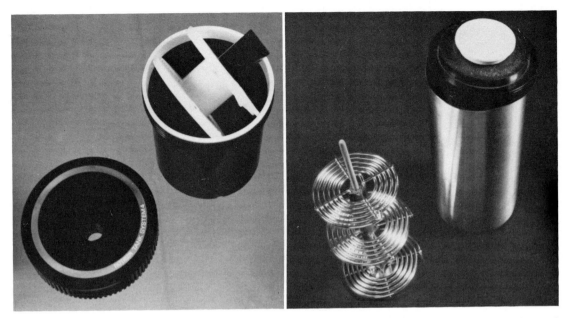

Figure 7.3.   Developing tanks for 35-mm film. **Left.** Plastic Paterson tank and reel. **Right.** Three-reel stainless steel Kinderman tank.

8. Suspend rack in a film-drying cabinet for 10–20 minutes.
9. When dry, put each negative in a polyethylene envelope and store in a filing cabinet.

Processing 35-mm Black and White Film

Chemicals and Equipment

- Paterson System 4 tank and 2 reels [Fig. 7.3 (left)]
or
  Stainless steel developing tank and 2 stainless steel reels (Kinderman tank, Catalog No. 3363; reels, catalog No. 3117) [Fig. 7.3 (right)].
- Developer appropriate to film used.
- Fixer.
- Kodak Photo-Flo.
- Graduated beaker, 1,000 mL.
- Scissors.
- Darkroom timer.

Method

Check whether film to be processed is panchromatic or orthochromatic; ortho films can be exposed to a specific type of safelight but panchromatic films, like Kodak Technical Pan 2415, must be handled in total darkness.

1. Make up developer and fixer according to manufacturers' instructions.
2. In total darkness or under appropriate safelight, open film cassette, remove film, cut a square end and load onto a developing reel. Place reel inside tank and put an empty reel on top (if using a 2–3 reel tank) to secure during agitation. If using

a metal tank, ensure that the reel cranks are pointing upwards. Fit lid securely. (If using a Peterson tank, fit the center spindle and place the reel in the tank with the flat side of the spindle down. Fit the lid securely.) Room lights can be switched on.

3. Start timer for time appropriate to the film being developed. Pour 300 mL (or the amount required to cover film) of developer into tank. Hold tank at an angle to facilitate air displacement and allow fluid to flow quickly into tank. Cap the lid. Agitate film by inverting tank three times within the first 10 seconds (tap the bottom of the tank sharply on a countertop to dislodge air bubbles from the film surface), then twice every 30 seconds thereafter.

4. Remove the cap and pour off developer. Rinse film by filling the tank to overflowing with 20°C water. Pour off.

5. Fill tank with fixer and agitate continuously for 2 minutes. Pour off fixer 1, replace with fixer 2 and agitate for 1 minute. Pour off. (Fixer may be used for 1 week. The next week fixer 2 becomes fixer 1 and fresh fixer is used in the second bath).

6. Remove tank lid and wash films in running water at 20°C for 20 minutes to remove excess fixer. Meanwhile, rinse tank lid and cap thoroughly with water and allow to air-dry.

7. Put film reel into a tank of dilute Kodak Photo-Flo for 30 seconds.

8. Unwind film strip from reel, attach film clips, and hang to dry in a film-drying cabinet for 10–20 minutes.

Notes on Processing 35-mm Film

- When first learning how to load a film onto a reel, practice with some waste film with room lights on.
- Paterson reels – engage film under ball bearings and pull forward approximately 3 cm. Grip the reel with both hands and, moving your hands in opposite directions, wind on the film.
- Metal reels – hold the reel in your left hand so that the wire crank is facing away from you. Fit the end of the film strip, emulsion side down, securely under the clip on the axle of the reel. Slightly pinch the edges of the film together with your right hand and turn the reel in an anti-clockwise direction. The film should wind onto the reel without resistance.
- Wash reels in water to remove Photo-Flo. Allow to dry completely before reloading with film.

PRINTING – GENERAL COMMENTS

The final product of so many hours of careful work is a high-quality print. Although it is best to work from a good negative, some shortcomings in the negative can be overcome when prints are made.

A positive print is made by projecting light from an enlarger through a negative onto photographic paper. Silver halide grains in the photographic emulsion of the paper are activated and the paper is developed, fixed, washed, and dried.

Although a test strip (made by exposing a sheet of photographic paper in strips for progressively longer periods of time) (Fig. 7.4A) can be used to determine correct exposure time, a little experience, combined with careful observation under consistent

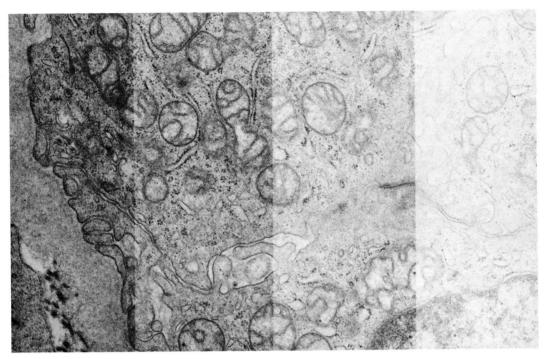

Figure 7.4. *Test prints used to determine correct exposure time and contrast.*
A. Exposure time ranged from 5 seconds (*right*) to 20 seconds.

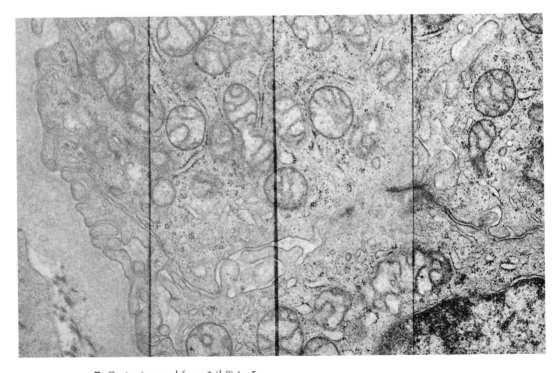

B. Contrast ranged from 0 (*left*) to 5.

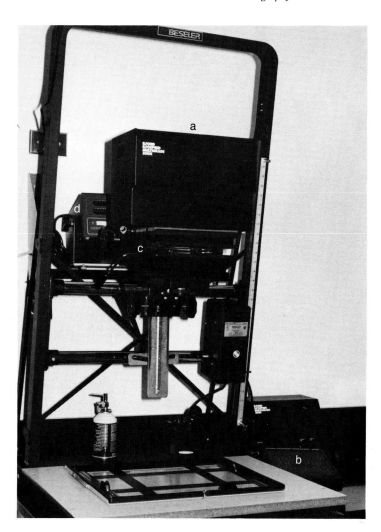

Figure 7.5. Semi-diffuser enlarger consisting of Ilford multigrade head which contains the light source and contrast filters (a); panel for filter and exposure time control (b); negative carrier (c); ionizing unit (d) (behind enlarger) for reducing static electricity and, therefore, dust.

safelight conditions, will soon enable even the beginner to judge both exposure time and contrast (Fig. 7.4B).

Uneven negative density can be corrected by "dodging" [using fingers or cardboard held above the paper to shadow the thinner (lighter) portion of the negative, thus exposing it for less time]. This is time consuming, wasteful of paper and rarely produces consistently good results. The time is better spent ensuring that the electron microscope is properly aligned and that the operator has even, well-cut, clean sections to micrograph. Although it is best to work with an ideal negative, one that is overexposed (more electrons) will produce a better print than one that is underexposed.

It is better to enlarge a negative than take a micrograph close to maximum magnification (e.g., a micrograph taken at 20,000 × and enlarged to 100,000 × will probably produce a print with higher contrast and better resolution than one taken at 100,000 ×). Negatives can be enlarged only to a certain point above which no more detail can be resolved; enlargement beyond the limit of resolution is called empty magnification, whether it occurs on a print or on a microscope (light or electron).

All darkrooms should be equipped with safelights appropriate to the films and papers being used. Kodak recommends checking safelights every 6 months since, as the filters fade, they begin to transmit the wavelengths of light they were designed to absorb (Kodak Publication No. K-4). To check safelights, place a coin on unexposed film or paper and leave under safelight conditions for 30 minutes. If, when developed, an image of the coin is visible, darkroom conditions are unsuitable, and corrective measures should be taken. All darkrooms should have a timer with a luminous dial that can be seen under safelight conditions.

ENLARGERS

There are three types of photographic enlargers available: a diffusion enlarger, a condenser enlarger, and a point light source condenser enlarger. A point source enlarger is considered the best choice for high-quality, high-resolution prints; a diffusion enlarger (Fig. 7.5) is not recommended. In practice, however, this is not necessarily found to be the case. Figure 7.6A was printed on a point source enlarger, Fig. 7.6B on a semi-diffuser enlarger. The negative was of correct density, and virtually no difference can be detected between the two prints.

When buying an enlarger, insist on a personal trial with your own negatives and compare types. The user is the best judge of whether a piece of equipment is easy and comfortable to handle and whether, for example, the quality gained in one instrument is sufficient to sacrifice the speed and efficiency of another. Price must also be a consideration.

Top quality lenses should be used regardless of cost. Lenses of correct focal length for the film should be used [e.g., if enlarging a 35-mm negative to 8 in. × 10 in. (200 mm × 250 mm), a 50-mm lens will be required; $3\frac{1}{4}$ in. × 4 in. (83 mm × 100 mm) negatives will require a 135-mm lens].

Enlargers must be properly maintained in order to produce high-quality prints. Lenses and condensers must be kept free of dust and fingerprints, which diffuse the light and thus reduce contrast, image sharpness, and print quality. The enlarger should be on a stable bench, free from vibrations. It might be advisable to install a constant-voltage transformer (voltage regulator), since a 10% reduction in voltage will produce a 30% reduction in light output (*Electron Microscopy and Photography*, Kodak Publication P-236).

Exposure of Negatives

1. Check that the enlarger lenses, stand, and easel are clean.
2. Position a negative, emulsion side down, in a clean carrier; ensure that it is flat and select paper size on easel.
3. Turn off room lights and turn on appropriate safelights.
4. Turn on enlarger and focus negative. With a point source enlarger, this will be done with the lens fully open. With other enlargers, stop down two f-stops from fully open. Use a magnifying focusing device (fine focusing an electron micrograph without one is virtually impossible).
5. Ensure that the negative is straight and fills the easel.
6. Determine proper exposure and contrast. The novice should do two test strips: one to determine exposure time (Fig. 7.4A), another to determine contrast (Fig. 7.4B).

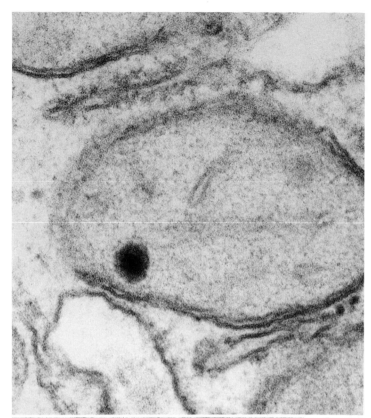

Figure 7.6. *Prints of a mitochondrion made from correctly exposed and developed negative.*

A. Print exposed on a point source enlarger.

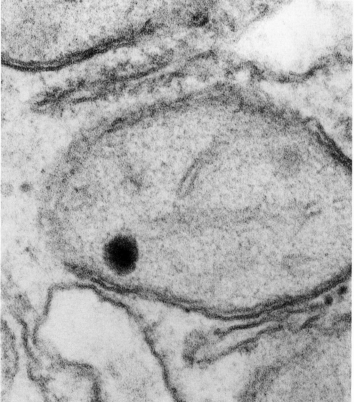

B. Print exposed on a semi-diffusion enlarger.

Experience, standard safelight conditions, and careful observation will soon enable even the beginner to determine these parameters by eye.

7. Turn off enlarger. Retrieve paper from light-tight package. Place paper, emulsion side up, on easel and expose.

Notes on Exposure of Negatives

- Use the first negative as a test and examine the resultant print carefully. You will remember (with experience) how the negative appeared on the easel and will be able to use it as a guide when setting exposure time and/or contrast on subsequent negatives. Remember, the print should show the information contained in the negative as completely and accurately as possible.

- If enhancement of some elements of a negative are desired, losses elsewhere are inevitable and unavoidable.

- Stopping a lens down two *f*-stops from fully open produces a flatter, sharper image. Stopping down too far will result in loss of definition.

- Flat negatives (those low in contrast) must be printed on hard (high-contrast) paper.

- Contrast must not be equated with resolution. In fact, the reverse is true; a low-contrast print has a longer grey scale, therefore more information, but is less pleasing to the eye. Working prints should be of low contrast; prints for publication or display, higher contrast.

- Extending development time will, to a point, produce more contrast, but too long in developer will produce fog and therefore less contrast. Underdevelopment of an over-exposed print reduces the contrast capabilities of a particular grade of paper. Optimum print quality results only when exposure and development times are exactly right.

- Exposure time will change if image size, contrast, or lens aperture opening are changed.

- Each sheet of paper should be labeled on the back with an identifying film and negative number. This should be done lightly in pencil ensuring that the imprint does not show on the emulsion side. Other information can be added later when room lights are on.

- Photographic paper is packaged so that the bottom one or two sheets are emulsion side up; all the rest are emulsion side down. Keep them that way to avoid getting fingerprints on the emulsion. If fingerprints are a problem, despite careful hand washing, wear gloves.

Tray Development of Prints

1. Cover exposed photographic paper with developer quickly and avoid trapping air bubbles on the emulsion. An image should appear in approximately 30 seconds and full development should be achieved in $1\frac{1}{2}$–2 minutes. Re-expose the negative, rather than over- or underdevelop the print.

2. Constantly rock developing tray with a gentle motion.

3. Handle prints by their edges only.

4. When developing is complete, pick up print by one corner, drain it a few seconds, and place it face up in stop bath for 5–10 seconds. Agitate continuously. This stops development and prepares the print for fixing.

5. Drain the print and place in fixer (hypo), face up, for 5–10 minutes. Agitate.
6. Rinse print briefly in water and put in hypo clearing agent; agitate for 2–3 minutes. Hypo clearing agent eliminates fixer and silver salts, resulting in faster and more thorough washing.
7. Wash 10–20 minutes in approximately 20°C water which is running fast enough to provide active agitation but does not fall directly on the print.
8. Bathe prints in print-flattening solution (to prevent curl) and dry on a clean, smooth surface.

Notes on Tray Development of Prints

- Fixer, stop bath, hypo clearing agent, and print-flattening solution can all be purchased from a local camera shop and are prepared according to manufacturers' instructions.
- Do not lift paper out of developer to check its progress. It may produce stains or safelight fog (which will reduce contrast and lower print quality). Safelight illumination is not bright enough to judge print quality in any case.
- It is best to alter exposure time rather than development time. Longer development can result in a stained print; shorter time, mottling or streaking.
- Prolonged time in either stop bath or hypo clearing agent may cause "water soak" (translucent areas on the print which may stain eventually).
- After 1–2 minutes in fixer, room lights may be turned on, the print examined and then returned to fixer for the required time. Fixer removes the "veiling" effect of unexposed and undeveloped portions of the emulsion, changing its appearance. Overfixing can cause fading.
- It is wise to use two fixing baths (placing the print half time in each). Discard the first fixer after 200 8 in. × 10 in. prints have been done. The second bath then becomes the first.
- A tray siphon (available from photographic supply stores) will convert any tray to an effective washing tray allowing water to circulate around prints and overflow into the sink. If hypo clearing agent has not been used, prints should wash for an hour. Without proper washing, they will yellow or fade in time.
- Dry prints on a clean, smooth surface. To minimize drying marks, wipe emulsion with a viscose sponge; it removes sediment and promotes faster drying. A commercial drier is recommended.
- When tray processing several prints at once, place them in developer in rapid succession and constantly bring the bottom print to the top (shuffle).
- Prints go into all solutions emulsion side up and should be agitated continuously.

PRINT PROCESSORS

Two types of print processor are available: (1) the stabilizing processor which produces a slightly damp, semipermanent print in 10–15 seconds and (2) the automatic processor which produces a permanent, dry print in 1–1.5 minutes. Both use developer-incorporated paper and both require complete accuracy when exposing the negative, since print development time cannot be controlled.

A stabilization print processor has an activator (an alkaline solution which combines with the developer incorporated in the surface of the paper) and a stabilizer (which neutralizes the activator and converts remaining silver halide to relatively stable, colorless compounds that remain in the paper). In time, the prints will yellow and fade. To make them permanent, these compounds must be removed with fixer and the prints washed and dried in the usual manner. Prints taken directly from the processor air-dry quickly and, if not exposed to light, high temperature, or humidity, may be stable for 2–3 years. These processors produce excellent quality prints in seconds. They are easy to use, requiring no special skill or training to operate them. They are also compact, require little bench space, and no special plumbing. Solutions are purchased ready to use and have an indefinite shelf life. For best results, the processor should be cleaned weekly or after an equivalent of 300 8 in. × 10 in. prints have been done. (CAUTION: Do not use aluminium foil near the activator solution, and do not mix activator and stabilizer, both result in the release of volatile gases that can cause respiratory tract and eye irritation.)

Automatic print processors make use of conventional processing (developing, fixing, washing, and drying). They require special plumbing and might require special ventilation if heat extraction is a problem. They are easy to maintain and produce high-quality permanent prints very rapidly.

Photographic Papers

Photographic papers fall into two main categories: (1) those that contain silver bromide in the emulsion (for use in standard tray development) and (2) those with developer incorporated in the gelatin as well (for use with print processors, although some may also be tray developed). A third type of paper is resin coated (RC), with developer incorporated; it can be tray or processor developed. All of these papers are available as either graded contrast (1–5) or polycontrast (requiring filters to alter contrast). Most come in more than one type of surface, but for transmission electron microscopy, a smooth, glossy surface gives the best image definition and tone reproduction.

Graded papers range from 1–5. Grades 1 and 2 are soft, have a long grey scale and are thus low in contrast. Grade 5 is hard, has a short grey scale, and is thus high in contrast (see Fig. 7.4B). A soft (low contrast) negative should be printed on hard paper and vice versa.

An advantage of polycontrast papers is that only one box of paper is required, eliminating the danger of little-used grades of paper expiring and therefore being wasted. Since polycontrast filters come in half grades, more control is possible than with single-grade papers.

Ilford produces a multigrade enlarger head (Fig. 7.5) with a control box that allows the selection of filters with the push of a button. It, combined with polycontrast paper and an automatic print processor will enable an experienced operator, with even negatives, to print a roll of 42 frames in less than an hour.

MEASUREMENT OF STRUCTURES ON MICROGRAPHS

The magnification of an electron micrograph is the product of the microscope magnification, camera factor, and photographic enlargement. The latter can be determined in

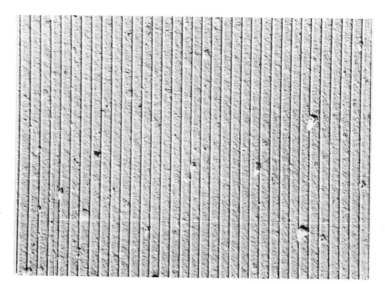

Figure 7.7. Calibration grid. Measurements are taken from one edge of first line to same relative edge of last line. The more lines measured, the greater the accuracy of the measurement.

two ways: (1) by measuring a structure on a negative and the same structure on the corresponding print or (2) by making a print of a transparent millimeter ruler placed at the same height as the original negative. Microscope magnification should be recorded when the micrograph is taken, and the camera factor will be given in the instruction manual supplied with the microscope.

For the identification of viruses, morphometry or other times when more accurate measurement is essential, latex spheres, which have very small size variation, can be applied directly to the specimen on the grid. Calibration grids (Fig. 7.7) are also available with instructions for their use. Care must be taken to measure from one edge of the first line to the same relative edge of the last line; the more lines measured, the more accurate the measurement. The grating in the following formula has 2160 lines per millimeter:

$$M = 2,160 \times X \div Y.$$

where M = magnification, X = total distance between limiting lines in millimeters, and Y = the number of spaces between the limiting lines.

Example:

$$M = 2,160 \times 92.5 \text{ mm} \div 6 = 33,300$$

Since the nanometers (nm) and the micron (μm) are the units of measurement used most often in electron microscopy and the millimeter (mm) the unit most often used to measure micrographs, the following relationships are useful:

$1 \text{ mm} = 1,000 \ (10^3) \ \mu\text{m}$

$1 \text{ mm} = 1,000,000 \ (10^6) \ \text{nm}$

$1 \ \mu\text{m} = 1,000 \ (10^3) \ \text{nm}$

Small objects may be measured using a magnifier incorporating a graticule. Larger objects may be measured with a millimeter ruler. A number of measurements should be made to determine an average, which when divided by the magnification will give true size.

Example: If the average measurement of an organelle on a negative is 1 mm and the microscope magnification is 5,000 ×, then

1 mm ÷ 5,000 (1,000 μm ÷ 5,000) = 0.2 μm (200 nm)

If measurements are taken from a print instead of the negative, the enlargement factor must also be included in the calculation.

Method for Putting Scale Bars on Micrographs

Choose a bar length that will be useful when comparing sizes of structures and organelles shown in the micrograph. Then, using the micrograph magnification, calculate its true length.

Example: to represent 0.1 μm on a micrograph of magnification 80,000 ×

Bar length = 0.1 × 80,000 ÷ 1,000 = 8 mm

Therefore, an 8-mm bar on the micrograph will represent 0.1 μm on the specimen.

Some microscopists prefer to use the same bar size, regardless of microscope magnification. In this case, the length that the bar represents will be different for each magnification and should be calculated as given in the next example.

Example: If the bar length = 10 mm and magnification = 50,000 ×, then the true size is given as

bar = 10 × 1,000 ÷ 50,000 = 0.2 μm (200 nm)

Example: If the bar length = 10 mm and magnification = 100,000 ×, then the true size is given as

bar = 10 × 1,000 ÷ 100,000 = 0.1 μm (100 nm)

Useful Size Relationships

Plasma membrane	0.0075 μm (7.5 nm)
Endoplasmic reticulum membrane	0.008 μm (8 nm)
Ribosome	0.015–0.023 μm (15–23 nm)
Centriole	0.150 μm (150 nm)
Mitochondria and some bacteria	0.2–0.5 μm (200–500 nm)
Nuclear pore	0.150 μm (150 nm)
Microtubule	0.018–0.020 μm (18–20 nm)
Glycogen	0.015–0.040 μm (15–40 nm)
Nuclear membrane	0.025–0.040 μm (25–40 nm)
Red blood cell	8 μm (0.008 mm)
Small secretory granules	0.100 μm (100 nm)
Actin filament	0.060 μm (60 nm)
Tonofilament	0.012 μm (12 nm)
Oocyte (mammalian)	100 μm (0.1 mm)
Liver cell	20 μm (0.02 mm)
DNA molecule	0.002 μm (2 nm)
For comparison:	
Wavelength of visible light	400–700 nm (0.4–0.7 μm)
Wavelength of 100 kv electron beam	0.0039 nm (0.0000039 μm)

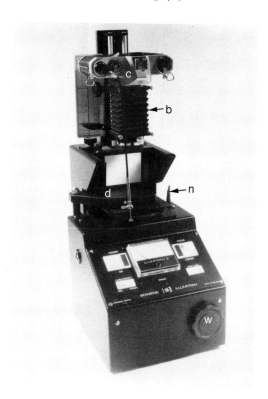

Figure 7.8. Bowens Illumitran. Camera (c); size ratio and focusing bellows (b); densitometer (d); negative carrier (n); wheel to adjust light intensity (w).

SLIDES FOR TEACHING

Transparencies from Direct Positive Film

The Kodak T-Max 100 direct positive kit (TMX 135-24, Catalog No. 830–5732) can be used to make high-quality positive black-and-white slides from continuous tone photographs, diagrams, X-ray films, and autoradiograms. The Kodak T-Max Film Developing Outfit (Catalog No. 195–4155), a copy stand, and camera body with cable release are required.

Transparencies from Black-and-White Negatives

Positive transparencies of 35-mm negatives can be made very easily using a Bowens Illumitran and a 35-mm camera (Fig. 7.8). The copy stand and camera mentioned above can be used to copy larger format negatives.

Copying 35-mm Negatives

1. Set up the Illumitran according to the manufacturer's instructions and "trim" using a negative of ideal density.
2. Place negative into holder, emulsion side up.
3. Load camera and bracket exposures.
4. Remove film and develop in D19.

Copying Larger Format Negatives

1. Load camera and attach to copy stand.
2. Place light box on baseboard of copy stand.
3. Place negative, emulsion side up, on the light box and mask peripheral light.
4. Adjust camera height to select desired area on the ground glass screen. Focus.
5. Bracket exposures.
6. Remove film and develop in D 19.
7. When dry, select optimal exposure from each group of three using a viewing box.
8. Place the grey half of a Gepe slide mount inside-up on the viewing box. Lay the selected frame emulsion-side-down and hook under the retaining metal strips on the mount ensuring that the image is properly aligned. To prevent the film from moving secure an edge to the inside of the mount with a piece of black tape. Lay the white half of the mount over the top of the grey half so that the film is sandwiched. Seal the mount using the Gepe Mounting Press (alternatively one can use finger pressure).
9. Place a small, circular, colored label on the lower left of the white half of the slide mount and identify the transparency with a marking pen. Transparencies are loaded in a projector carousel so that the colored dot is in the upper right corner of the slide (the grey half of the slide mount and emulsion side of the negative face the screen).

Copying Micrographs

When a black-and-white print is required of a micrograph to which lettering has been added or of a montage consisting of several micrographs, Kodak Technical Pan Film 2415, 135-36, Technidol LC Developer or D 19 and a copy stand and camera will produce a negative (from which prints can be made) with resolution and contrast similar to those of the original.

REFERENCES

Eastman Kodak Company. (1973). *Electron Microscopy and Photography*. Publication No. P-236, Cat. 179, 7703.

Eastman Kodak Company. (1985). *Kodak Electron Imaging Plates*. Publication No. N-921.

Guerney, R. W., & Mott, N. F. (1938). *Proceedings of the Royal Society, London, Ser. A.* **164**: 151.

SUGGESTED PUBLICATIONS FROM THE EASTMAN KODAK COMPANY*

Index to Kodak Information. Kodak Publication No. L-5. [A complete list of Kodak information books and pamphlets.]

Tech Bits. [Brings imaging techniques to scientists and engineers; published three times a year.]

How Safe is Your Safelight? Kodak Publication No. K-4.

*For more information, telephone the Kodak Information Center, Scientific Imaging Staff, 1-800-242-2424, Ext. 12 (U.S.A.) or 1-800-387-8773 (Canada). Or write to Eastman Kodak Company, Dept. L-5, 175 Humboldt Street, Rochester, NY 14610-1099.

Practical Processing in Black and White Photography. (1976). Kodak Publication No. P-229, Cat. 146, 1150.

Electron Micrography – Using Electrons Effectively. (1977). Kodak Pamphlets Nos. P-317 & M6-121.

Kodak Products for Electron Micrography. (1987). Kodak Publication No. N-923.

Kodak Scientific Imaging Products. (1989). Kodak Publication No. L-10.

• *Appendix*

DEFINITIONS (*HANDBOOK OF CHEMISTRY AND PHYSICS 1988–1989*)

Acids are hydrogen-containing substances that dissociate on solution in water to produce one or more hydrogen ions. An acid is anything that can attach itself to something with an unshared pair of electrons.

Bases are substances that dissociate on solution in water to produce one or more hydroxyl ions. A base is anything that has an unshared pair of electrons.

A *salt* is any substance that yields ions other than hydrogen or hydroxyl ions. A salt is obtained by replacing the hydrogen of an acid by a metal.

A *normal solution* (a former expression of concentration) is the number of gram equivalents of the substance per liter of solution.

A *molal solution* contains one gram molecular weight of the solute per kilogram of solution (mol/kg).

A *molar solution* contains one gram molecular weight (M – relative molecular mass) of the solute per liter of solution (mol/L).

The *molecular weight* is the sum of the atomic weights (A_r – relative atomic mass) of all the atoms in a molecule.

A *molecule* is the smallest unit quantity of matter that can exist by itself and retain all the properties of the original substance.

Concentration refers to the amount in weight, moles, or equivalents contained in a unit volume.

Daltons refers to atomic mass units.

Gram atomic weight is the mass in grams numerically equal to the atomic weight.

Gram equivalent weight is the weight of a substance displacing or otherwise reacting with 1.008 grams of hydrogen or combining with 0.5 of a gram atomic weight (8.00 grams) of oxygen.

Gram molecular weight or *gram molecule* is a mass in grams of a substance numerically equal to its molecular weight (gram mole).

Gram mole, gram formula weight, or *gram equivalent weight* is the mass in grams numerically equal to the molecular weight, formula weight, or chemical equivalent, respectively.

Oxidation is any process that increases the proportion of oxygen- or acid-forming element or radical in a compound.

The *index of refraction* (*n*) for any substance is the ratio of the velocity of light in a vacuum to its velocity in the substance and varies with the wave length of the refracted light. It is also the sine of the angle of refraction.

The *numerical aperture* is the sine of half the angular aperture and is used as a measure of the optical power of an objective lens.

The *resolving power* of a microscope is the minimum distance at which two objects can be recognized separately when viewed through the instrument.

Focal length is the distance between the optical center of a lens and its focus.

Focal plane is a plane parallel to the plane of a lens and passing through the focus.

Focus (focal point) is that point at which parallel rays of light meet after refraction by a lens.

Astigmatism is the ability of a lens to magnify equally in all planes.

Chromatic aberration is due to a difference in index of refraction for different wavelengths. Light of various wavelengths from the same source cannot be focused at a point by a simple lens.

Achromatic is the term applied to lenses to signify their more or less complete correction for chromatic aberration.

Spherical aberration occurs when large lenses are used, and the light divergent from a point source is not exactly focused at a point.

An *anode* is the electrode at which oxidation occurs. It is the electrode toward which anions travel due to the electrical potential.

A *cathode* is the electrode at which reduction occurs. In a vacuum tube, it is the electrode from which electrons are liberated.

Diffraction is that phenomenon produced by the spreading of waves around and past obstacles that are comparable in size to their wavelength.

Elastic collision is a collision between two particles in which no change occurs in the internal energy of the particles or in the sum of their kinetic energies (billiard-ball collision).

An *electron* is a small particle having a unit negative electrical charge, a small mass, and a small diameter.

Electron volt (eV) is the energy required by any charged particle carrying unit electronic charge while it falls through a potential difference of 1 volt.

Fringe is the locus of maximum constructive interference (light fringe) or destructive interference regions in a space where two or more coherent waves intersect. It can be in two or three dimensions.

Fresnel is a measure of frequency defined as equal to 10^{12} cycles per second.

Hysteresis is the magnetization of a sample of iron or steel due to a magnetic field which, made to vary through a cycle of values, lags behind the field.

Fluorescence is the property of emitting radiation as the result of absorption of radiation from some other source. The emitted radiation persists only as long as the substance is subjected to radiation, which may be either electrified particles or waves. The fluorescent radiation generally has a longer wavelength than that of the absorbed radiation.

Torr is the unit used to replace the English unit mmHg and is 1/760 of a standard atmosphere or 1333.22 microbars.

International System of Units [Système International (SI)] is a metric system of units founded on seven base units (meter, second, kilogram, mole, ampere, Kelvin degrees, and candela). Also called the *MSKA System.* Any physical quantity can be expressed by appropriate combinations of these units.

Physical Quantity	Base Unit	Symbol
Length	meter	m
Mass	kilogram	kg
Time	second	s
Amount of substance	mole	mol
Thermodynamic temperature	Kelvin	K
Electric current	ampere	A
Luminous intensity	candela	cd

Prefix Names of Multiples and Submultiples of Units

Factor by which Unit is Multiplied	Prefix	Symbol
10^{12}	tera	T
10^{9}	giga	G
10^{6}	mega	M
10^{3}	kilo	k
10^{2}	hecto	h
10	deka	da
10^{-1}	deci	d
10^{-2}	centi	c
10^{-3}	milli	m
10^{-6}	micro	μ
10^{-9}	nano	n
10^{-12}	pico	p

Prefixes

Deci — 1/10 or 10^{-1}
Kilo — 1,000 or 10^{3}
Micro — 1/1,000,000 or 10^{-6}
Milli — 1/1,000 or 10^{-3}

CONVERSION FACTORS

To Convert From	To	Multiply By
Ångström units	Nanometers	1×10^{-1}
	Micrometers	1×10^{-4}
	Centimeters	1×10^{-8}
Nanometers	Ångström units	1×10^{1}
	Micrometers	1×10^{-3}
	Millimeters	1×10^{-6}
	Centimeters	1×10^{-7}
Micrometers	Ångström units	1×10^{4}
	Nanometers	1×10^{3}
	Millimeter	1×10^{-3}

	Centimeters	1×10^{-4}
	Meters	1×10^{-6}
Centimeters	Ångström units	1×10^{8}
	Nanometers	1×10^{7}
	Micrometers	1×10^{4}
	Millimeters	1×10^{1}
	Meters	1×10^{-2}
Meters	Ångström units	1×10^{10}
	Nanometers	1×10^{9}
	Micrometers	1×10^{6}
	Millimeters	1×10^{3}
	Centimeters	1×10^{2}
Cubic millimeters	Cubic centimeters	1×10^{-3}
	Cubic meters	1×10^{-9}
Cubic centimeters	Cubic meters	1×10^{-6}
Cubic meters	Cubic centimeters	1×10^{6}

Equivalents and Conversion Factors

1 meter	$= 39.37$ in.
1 kilogram	$= 2.2$ lb
1 inch	$= 2.54$ cm $= 25.4$ mm
1 pound	$= 454$ g
1 grain	$= 65$ mg
1 fluid ounce	$= 28.41$ mL
1 pint	$= 568$ mL
1 gallon (imp)	$= 4.546$ liter

Thermometric Conversions

To convert from degrees centigrade to degrees Fahrenheit: Multiply by 9/5 or 1.8 and add 32. To convert from degrees Fahrenheit to degrees centigrade: Subtract 32 and multiply by 5/9 or 0.555.

SI Units – Equivalents

SI Units	*Former Equivalents*
1 pascal (Pa)	1 newton/m^{2}
100 pascals	1 millibar (mbar)
133 pascals	1 torr $= 1$ mmHg
	$= 1.33$ mbar
	760 torr $= 760$ mmHg
	$= 1$ atmosphere
	$= 1.013$ bar
10^{5} pascals	1 bar

In the electron microscope column the normal operating vacuum should be between 10^{-2} and 10^{-3} Pa (10^{-4} and 10^{-5} torr).

Symbol	SI Units
nm	nanometer
μm	micrometer
mm	millimeter
g	gram
mg	milligram
°C	degrees centigrade
d	day
h	hour
min	minute
s	second
L	liter
mL	milliliter
mol/L	moles/liter

COMPENSATING FOR HYDRATED STATES OF A CHEMICAL

If a method calls for an anhydrous chemical and the one available is a hydrated compound, one must compensate for the water. For example, if the method calls for 3 g of anhydrous sodium carbonate (Na_2CO_3, formula weight 106.00) and the reagent available is hydrated ($Na_2CO_3 \cdot 10H_2O$, formula weight 286.16), one would divide the formula weight of the hydrated compound by the formula weight of the anhydrous compound and multiply by the amount of anhydrous compound called for in the method:

286.16 g $\div$ 106 $\times$ 3 g = 8.099 g of the hydrated compound

The formula weight of water is 18.01.

DILUTIONS

Preparation of percent solutions from solids (w/v)
  Example: 1% toluidine blue.
  Dissolve 1 g toluidine blue in sufficient solvent and make up to 100 mL.
Preparation of percent solutions from liquids (v/v)
  Example: 3% hydrochloric acid.
  Add 3 mL of concentrated hydrochloric acid to distilled water and make up to 100 mL.
Dilution of already-diluted solutions
  Example 1: 5% hydrochloric acid from 36% hydrochloric acid. Pipette 5 mL of 36% hydrochloric acid into distilled water and make up to 36 mL. The percentage on hand (36) minus the percentage required (5) equals the volume of diluant to be added (31).
  Example 2: 100 mL of 30% alcohol from 95% alcohol.
  30 $\times$ 100 $\div$ 95 = 31.5 mL of 95% alcohol required
Percentage required $\times$ volume required $\div$ percentage on hand equals the amount of

chemical on hand that must be diluted to the required volume. Although these are quick, practical examples, the general rule in all dilutions is $C_1V_1 = C_2V_2$ [concentration on hand (C_1) × volume (V_1) on hand = concentration required (C_2) × volume (V_2) required].

Strengthening a solution by addition of a solid

Example: To make 100 mL of a 25% solution from a 20% solution.

Take 90 mL (for example) of the 20% solution. It contains:

$$90 \div 100 \times 20\,g = 18\,g$$

Add 7 g of the pure substance $(18 + 7 = 25)$ and dilute to 100 mL.

NORMAL SOLUTIONS

A normal solution is one gram molecular weight of the dissolved substance divided by the hydrogen equivalent (one gram equivalent) of the substance per liter of solution.

Example 1 (Solids): Normal sodium hydroxide (NaOH).

Relative molecular mass Na = 23   Valency = 1

O = 16

H = 1

Total = 40

Equivalent weight = 40 ÷ 1 = 40

1 N NaOH = 40 g NaOH dissolved in distilled water to a total volume of 1,000 mL.

Example 2 (Liquids): Normal hydrochloric acid (HCl)

The formula is: relative molecular mass ÷ (valency × specific gravity × concentration)

Relative molecular mass = 36.4

Valency   = 1

Specific gravity   = 1.18

Concentration   = 36%

$36.4 \div (1 \times 1.18 \times 0.36) = 85.7\,ml$

1 N HCl = 85.7 mL concentrated HCl diluted to 1,000 mL.

Example 3: Normal sulfuric acid (H_2SO_4)

Relative molecular mass   = 98.08

Valency   = 2

Specific gravity   = 1.84

Concentration   = 98%

$98.08 \div (2 \times 1.835 \times 0.98) = 27.2\,mL$

1 N H_2SO_4 = 27.2 mL concentrated H_2SO_4 diluted to 1,000 mL.

All solutions are made up in distilled water unless the method states otherwise.

APPROXIMATE ATOMIC WEIGHTS OF SOME COMMON ELEMENTS

Hydrogen (H) =	1	Magnesium (Mg) =	24
Carbon (C)	= 12	Phophorus (P)	= 31
Nitrogen (N)	= 14	Chlorine (Cl)	= 35.5
Oxygen (O)	= 16	Potassium (K)	= 39
Sodium (Na)	= 23	Calcium (Ca)	= 40

ELECTRON MICROSCOPY ACCESSION LOG

University Hospital
Owned and Operated by
London Health Association

SPEC. NO.	DATE	NAME OF PATIENT	SPECIMEN	HOSPITAL	COMMENTS	BLOCKS	T.B.	Ag	GRIDS	BOX	FILM	PRINTS	TRANS
90.S 425	4.2.90	Smith John	renal bx	UH	Bx Root	6	3	3	3	29356 A2, A3, A4	1124 (3-12)	10	4

8401.1131 5 (66)1/04/90)10 Rev NS

Figure A.1. Electron microscopy accession log.

SOME USEFUL pH VALUES

Hydrochloric acid, 1 N	0.1
Hydrochloric acid, 0.1 N	1.1
Hydrochloric acid, 0.01 N	2.0
Acetic acid, 1 N	2.4
Acetic acid, 0.1 N	2.9
Sodium hydroxide, 1 N	14.0
Sodium hydroxide, 0.1 N	13.0
Potassium hydroxide, 1 N	14.0
Potassium hydroxide, 0.1 N	13.0
Potassium hydroxide, 0.01 N	12.0
Calcium carbonate, saturated solution	9.4
Borax, 0.1 N	9.2
Blood plasma (human)	7.3–7.5
Spinal fluid (human)	7.3–7.5
Saliva (human)	6.5–7.5
Gastric contents (human)	1.0–3.0
Duodenal contents (human)	4.8–8.2
Feces (human)	4.6–8.4
Urine (human)	4.8–8.4

RECORD KEEPING

Every laboratory should keep accurate records of all specimens processed. Each page of the accession log (see previous page) provides easy access to information on 15 specimens. It has space for comments (e.g., whether a specimen was in an unsuitable fixative before we received it, whether we went to the biopsy, whether the specimen was taken from paraffin, etc). We record how many pieces of tissue were embedded (Blocks); how many semithin sections were cut (TB); whether we did a silver stain on the sections (Ag); how many grids were cut (Grids); where those grids are stored (Box); and, if micrographs were taken, how many and the film number. We also record whether transparancies (Trans) were done since the author's laboratory is in a teaching hospital and transparencies are frequently required for hospital rounds.

REFERENCES

Handbook of Chemistry and Physics (1988–89). 69th ed. Boca Raton: CRC Press.
Compendium of Pharmaceuticals and Specialities (1989). 24th ed. Ottawa: Canadian Pharmaceutical Association.
The Merck Manual (1987). 15th ed. Rahway: Merck Sharp & Dohme Research Laboratories.

SUPPLIERS

The following is a partial list of suppliers for the chemicals and equipment used in transmission electron microscopy.

Chapter 1 – Fixation

Glutaraldehyde, Osmium, and General Supplies

- Agar Aids, 66a Cambridge Road, Stanstead, Essex CM24 8DA, England.
- International Instruments, Ltd., Oosman Chamber, P.O. Box 15040, Karachi-3, Pakistan. (Agents for Agar Aids)
- Analychem Corp., Ltd., 7321 Victoria Park Avenue, Unit 16, Markham, Ontario L3R 2Z8, Canada. (Agents for Polysciences)
- Allied Fisher Scientific, Ltd., 1200 Denison Street, Unionville, Ontario L3R 8G6, Canada.
- Baxter Corp., Canlab Division, 2390 Argentia Road, Mississauga, Ontario L5N 3P1, Canada.
- Baxter Healthcare Corp., 1 Baxter Parkway, Deerfield, IL 60015. (Also in Belgium, Turkey, UK, Equador, Kuwait, Mexico, and Thailand)
- Better Equipment for Electron Microscopy (BEEM), Inc., P.O. Box 132, Jerome Avenue Station, Bronx, NY 10468.
- Ebtec Corp., 120 Shoemaker Lane, P.O. Box 465, Agawam, MA 01001. (Agents for TAAB)
- Electron Microscopy Sciences, Box 251, Fort Washington, PA 19034.
- Electron Microscope Aids, Chestnut House, 72 Dragon Road, Winterbourne, Bristol, England.
- EM Industries, 5 Skyline Drive, Hawthorne, NY 10532. (Distributors in 30 countries)
- Ernest F. Fullam (EFFA), Inc., P.O. Box 444, Schenectady, NY 12301. International Office: 900 Albany Shaker Road, Latham, NY 12110. [Distributors in Canada, Continental Europe (France), UK, and Japan]
- Graticules, Ltd., Sovereign Way, Botany Trading Estate, Tonbridge, Kent TN9 1RN, England. (Agents for Ernest F. Fullam)
- Ingram & Bell Scientific, 20 Bond Avenue, Don Mills, Ontario M3B 1L9, Canada.
- Johnson & Matthey Chemicals, Ltd., Orchard Road, Royston, Hertfordshire SG8 5HE, England.
- Ladd Research Industries, Inc., P.O. Box 901, Burlington, VT 05402.
- Marivac Services, 5821 Russell Street, Halifax, Nova Scotia B3K 1X5, Canada. (Agents for TAAB)
- Pelco International, P.O. Box 510, Tustin, CA 92681
- Polaron Equipment, Ltd., 60/62 Greenhill Crescent, Holywell Estate, Watford, Hertfordshire WD1 8XG, England. (Agents for Polysciences)
- Polysciences, Inc., 400 Valley Road, Warrington, PA 18976. [Distributors in Argentina, Australia, Brazil, Canada, Chile, Colombia, UK, Europe (Germany), India, Indonesia, Israel, Japan, Korea, Kuwait, Malaysia, Mexico, New Zealand, Philippines, Singapore, Taiwan, Thailand, United Arab Emirates, and Venezuela]
- Soquelec Telecommunications, Ltd., 5757 Cavendish Blvd. #101, Montreal, Quebec H4W 2W8, Canada.
- SPI Supplies, P.O. Box 342, West Chester, PA 19380.
- TAAB Laboratory Equipment, Ltd., 3 Minerva House, Calleva Industrial Park, Aldermaston, Berkshire RG7 4QW, England.

Chemicals

- BDH Inc., 350 Evans Avenue, Toronto, Ontario, M8Z 1K5, Canada.
- BDH Ltd., Poole, England.
- Crown Scientific, Private Mail Bag 6, Auburn, NSW 2144, Australia. (Agents for BDH)
- J. T. Baker, Inc., 222 Red School Lane, Phillipsburg, NJ 08865.
- J. T. Baker Canada, P.O. Box 355, Station A, Toronto, Ontario M5W 1C5, Canada.
- J. T. Baker UK, Hayes Gate House, Hayes, Middlesex UB4 0JD, England. (Distributors in Holland, Germany, Mexico, Venezuela, and Singapore)
- Mallinckrodt Canada, Inc., 600 Dalmar Avenue, Pointe Claire, Quebec H9R 4A8, Canada.
- Mallinckrodt, Inc., 675 McDonnell Blvd., St. Louis, MO 63134. (Agents in New Zealand, Indonesia, Ireland, Korea, Norway, Peru, and Sweden)
- E. Merck, Darmstadt, Germany.

Razor Blades

- Pelco International.
- Ladd Research Industries, Inc.
- Ingram & Bell Ltd. (Distributor for Ladd).

Glassware

- Fisher Scientific, 112, Ch Colonnade Road, Nepean, Ontario K2E 7L6, Canada.
- Johns Scientific Inc., 175 Hanson Street, Toronto, Ontario M4C 1A7, Canada.

Iris Forceps

- Sklar Manufacturing Co., Inc. J. 201 Carter Drive, West Chester, PA 19380.

Microtiter Plates

- Becton Dickenson and Company, 2 Bridgewater Lane, Lincoln Park, NJ 07035.
- Becton Dickenson Canada Inc., 2464 S. Sheridan Way, Mississauga, Ontario L5J 2M8, Canada.

Perforated Sheets of Labels

- Lipshaw Corp., 7446 Central Avenue, Detroit, MI 48210.
- Ingram & Bell Scientific (Canada).
- Schuco International London, Ltd., Halliwick Court Place, Woodhouse Road, London NI2 0NE, England. (Agents for Lipshaw)

Dissecting Microscopes

- Carl Zeiss Inc., One Zeiss Drive, Thornwood, NY 10594.
- Olympus Corp., 4 Nevada Drive, Lake Success, NY 11042.
- Leica Canada, Ltd., 513 McNicoll Avenue, Willowdale, Ontario M2H 2C9, Canada.
- Leica USA, 24 Link Drive, Rockleigh, NJ 07647.

Rotator

- Fisher Scientific, Ltd.

Dental Wax

- Dentsply/York Division, Dentsply International Inc., 570 West College Avenue, York, PA 17405-0872. (Agents in Philippines, Sweden, Bermuda, Colombia, Dominican Republic, Equador, Panama, Saudi Arabia, and UK)

Detergent

- Teepol – Available from most chemical supply houses.
- Liqui-nox – Alconox, Inc., New York, NY 10003.

Parafilm

- American Can Co., American Lane, Greenwich, CT 06836.

Chapter 2 – Dehydration and Embedding

Balances

- General suppliers – See Chapter 1

Hypodermic Syringes

- Beckton Dickenson, Stanley St., Rutherford, NJ 07070. (Agents in Austria, Puerto Rico, Canada, Denmark, Kuwait, Mexico, Qatar, Saudi Arabia, and UK)

Molecular Sieve

- BDH

Plastisolve

- Allied Fisher Scientific

Polyethylene Transfer Pipettes

- Allied Fisher Scientific

Ethyl Alcohol, Absolute

- Midwest Grain Products, South Front St., P.O. Box 1069, Pekin, IL 61555.

Chapter 3 – Cutting

Diamond Knives

- Dia-Tech Corporation, 6408 Clinton Highway, Suite 5, Knoxville, TN.
- Marivac Services (Canadian distributor for Dia-Tech).

- EM Supplies Division, 12 Middlesex Rd., P.O. Box 285, Chestnut Hill, MA 02167.
- Hacker Instruments Inc., Box 657, Fairfield, NJ 07007.
- Leica (Reichert-Jung), Box 123, Buffalo, NY 14240.
- Micro Engineering Inc. Route 2, Box 474, Huntsville, TX 77340.
- TAAB Laboratories
- Agar Aids

Knife Makers – LKB, TAAB-Pyper

- Leica Canada
- Leica USA
- TAAB Laboratories
- Agar Aids
- Marivac Services

Plastic Tape – 3M

- Ernest F Fullman (and agents)
- 3M Sepei Ltd., 30 Abdel Rahim Sabri St., Dokki, Cairo, Egypt.
- Petsiavas N.S.A., 11 Nicodemou St., G-105 Athens, Greece.
- Sewart Cottrell & Co. J., 15-17 Charlotte St., London WP1 2AA, England.

Ultramicrotomes – Reichert-Jung, LKB

- Leica Canada
- Leica USA

Ultramicrotomes – Sorvall

- DuPont (UK), Ltd., Instrument Products Division, Wilbury House, Wilbury Way, Hitchin, Hertfordshire SG4 0UR, England.
- E. I. DuPont de Nemours & Co, Inc., Instrument Products Division, 1007 Market St., Wilmington, DE 19898.

Butterfly Scalp Vein Infusion Sets

- Sherwood Medical Co., 1831 Olive St., St. Louis, MO 63103.
- Beckton Dickenson Division, Medfusion Systems, Inc., 3070 Business Park Dr., Norcross, GA 30071.

Micropore Filters

- Gelman Sciences Inc., 2535 de Miniac, Montreal, Quebec H4S 1E5, Canada.
- Gelman Sciences Inc., 600 S. Wagner Rd., Ann Arbor, MI 48106. (Agents in Pakistan, Puerto Rico, and Brazil)
- Millipore Corp., 80 Ashby Rd., Bedford, MA 01730. (Agents in Qatar, Indonesia, and Portugal)
- Millipore (Canada) Ltd., 3688 Nashua Drive, Mississauga, Ontario L4N 1M5, Canada.

- Nuclepore Corp., 7035 Commerce Circle, Pleasanton, CA 94566.
- Nuclepore Canada Inc., 30 Fordhouse Blvd., Toronto, Ontario M8Z 5M3, Canada.

Specimen Support Grids

- Mason & Morton, Ltd., M & M House, Frogmore Road, Hemel Hempstead, Hertfordshire HP3 9RW, England.
- Smethurst Highlight Ltd., 420 Chorley New Road, Bolton, Lancashire BL1 5BA, England.
- Gilder Grids, 23 Mcfarlane Road, London WI2 7JY, England.
- Veco Ltd., 36 Essendene Road, Caterham, Surrey CR3 5PA, England.

Polypropylene Tubes (Polyallomer)

- Beckman Instruments Inc., 2500 Harbor Blvd., P.O. Box 3100, Fullerton, CA 92634. (Agents in Bolivia, Belgium, Aruba, Egypt, Israel, Kuwait, Saudi Arabia, Spain, Uruguay, and Colombia).

Static Eliminators

- NRD Inc. (Nucleospot), 2937 Alt Blvd. North, Grand Island, NY 14072.
- CanadaWide Scientific Ltd. (Staticmaster Ionizing Unit), 1230 Old Innes Rd., Unit 414, Ottawa, Ontario K1B 3V3, Canada.

Chapter 4 – Immunoelectron Microscopy

General Suppliers

- Cedar Lane Laboratories Ltd., 5516-8th Line, RR2 Hornby, Ontario L0P 1E0, Canada.
- Accurate Chemicals & Scientific Corp., 300 Shames Drive, Westbury, NY 11590.
- Sera-Lab Ltd., Crawley Down, Sussex RH10 4LL, England. (Both distributors for Cedar Lane; distributors also in Germany, France, Switzerland, Holland, Sweden, Italy, Austria, Israel, Japan, Norway, and Taiwan)
- Miles Scientific, 30W475 N. Aurora Rd., Naperville, IL 60566.
- Miles Scientific, 7 Belfield Rd., Rexdale, Ontario M9W 1G6, Canada.
- Miles Scientific, Stoke Poges, Slough SI2 4LY, England. (Agents also in Australia, Austria, France, Germany, Italy, Japan, Netherlands, Spain, and the Caribbean)
- E.Y Laboratories Inc., 127 N. Amphlett Blvd., San Mateo, CA 94401.
- Bioprocessing Ltd., Consett 1 Industrial Estate, Medomosley Rd., Consett, Co. Durham DH8 6TJ, England. (Distributor for E.Y; distributors also in Australia, France, Holland, Japan, Spain, Sweden, Switzerland, and Germany)
- Difco Laboratories Inc., P.O. Box 1058-A, Detroit, MI 48232 and Sao Paulo, Brazil.
- Janssen Life Sciences Products, Turnhoutseweg 30, B-2340 Beerse, Belgium.
- Janseen Life Sciences Products, 40 Kingsbridge Rd., Piscataway, NJ 08854.
- Inter Medico, Suite 307, 150 Consumers Rd., Willowdale, Ontario M2J 1P9, Canada.

Rabbit Anti-human Immunoglobulin

- Dako Corp., 22 North Milpas St., Santa Barbara, CA 93103.

Micropipettor and Tips – Eppendorf

- Thomas Scientific, 99 High Hill Rd., P.O. Box 99, Swedesboro, NJ 08085.
- Crown Scientific, Australia., (Distributors for Eppendorf also in Austria, Luxembourg, Norway, and Belgium)
- Allied Fisher Scientific.
- Baxter.

Tween 20

- J. T. Baker Chemical.
- Allied Fisher Scientific.

Chapter 5 – Special Methods

Thermanox Coverslips

- Nunc Inc., Napierville, IL.
- Gibco/BRL Canada, 2270 Industrial St., Burlington, Ontario L7P 1A1, Canada.

Vacutainer Tubes

- Beckton Dickenson.

Centrifuge

- Beckman Instruments Inc., 1050 Page Mill Rd., Palo Alto, CA 94304.

Egg Albumin

- Sigma Diagnostics, 3050 Spruce St., St. Louis, MO 63103

Bovine Serum Albumin

- Sigma Diagnostics
- Interlab Distribuidora de Productos Cient, Rua Luis Goes 853, 04043 Sao Paulo, Brazil.

Agar

- Difco.

• *Index*